List of Figures and Tables

Contents

1.0 Introduction

1.1 Preface

Low voltage (LV) and High Voltage (HV) electrical circuits have varying types of protection relays, circuit breakers and fuses for both safety and damage limitation purposes. All of which require maintenance to ensure continued safe and reliable service.

Original Equipment Manufacturers (OEM) and numerous technical authorities have written textbooks, manuals and papers regarding switchgear. However, much of the information required for electrical fitters, engineers and maintenance technicians has to be extracted from different sources and gained through experience.

1.2 Aim

The aim of this guidance document is to provide technicians, students and engineers with an overall appreciation of typical maintenance practices for both switchgear and protection.

1.3 Objectives

The main objectives of this paper are to:

- Understand the difference between, effects of and protective measures for overload, overcurrent and earth leakage faults.
- Learn the key components of a protection system including its characteristics.
- An awareness different types of circuit breakers
- The operating properties of a contactor and circuit breaker
- Provide basic essentials and guidelines for protection system maintenance
- Sample worked instructions for the maintenance of both electromechanical and electronic relay protection relays.
- Sample worked instructions of real life maintenance practices

1.4 Scope

The circuit breakers covered within this document are limited to a maximum operating voltage of 11,000 volts.

The examples and worked instructions provided will typically be for motor circuits.

The sample techniques and practices are a generic approach; specific instructions should be sought from the manufacturer's guidelines.

The various suggested maintenance methods are by no means exhaustive and are limited to the author's own knowledge and experience.

Explanations of various system components are in most cases abridged and not guaranteed to be 100% accurate but are suitable for the aims of this document.

A list of recommended further reading will be provided at the end.

2.0 Types of Fault

2.1 Introduction

An electrical fault can be defined as an abnormal operating condition or the potential to cause damage or harm. For the purposes of this document we will be looking at overload, overcurrent and earth fault conditions.

2.2 Conductor Revision

From the figure 1 below if a smaller diameter conductor is placed between the larger conductors it will become warmer than the larger conductor if a current is passed through it. A smaller diameter offers more resistance which therefore will produce heat.

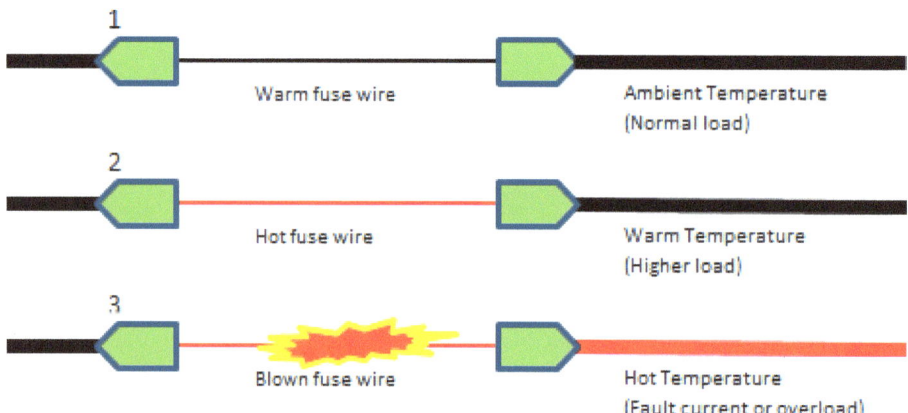

Figure 1: Conductor revision

If the current is increased the larger conductor will get warm whilst the smaller will get hot. A fault or overload current is much larger than the normal current and eventually the smaller diameter conductor will melt. Other protective devices work on the same principle although they are not self-destructive and can be reset.

2.3 Overload

An overload situation might be where a motor is driving a fan via a belt and pulley system. If the fan is drawing air through a filter and the filter becomes excessively clogged then the motor will have to work harder to turn the fan, thus increasing the normal running load. An increase in load, results in an increase in running current.

To summarise, an overload on an electrical circuit is a slower event and might not lead to immediate failure. Circuit components, protective devices (fuse or circuit breaker) and cables are normally designed to withstand at least 10% above their rated operating current for a certain amount of time. For example a 100A circuit being overloaded at

110A will probably continue to operate for approximately one hour with no adverse effects, although circuit components will have a reduced lifespan.

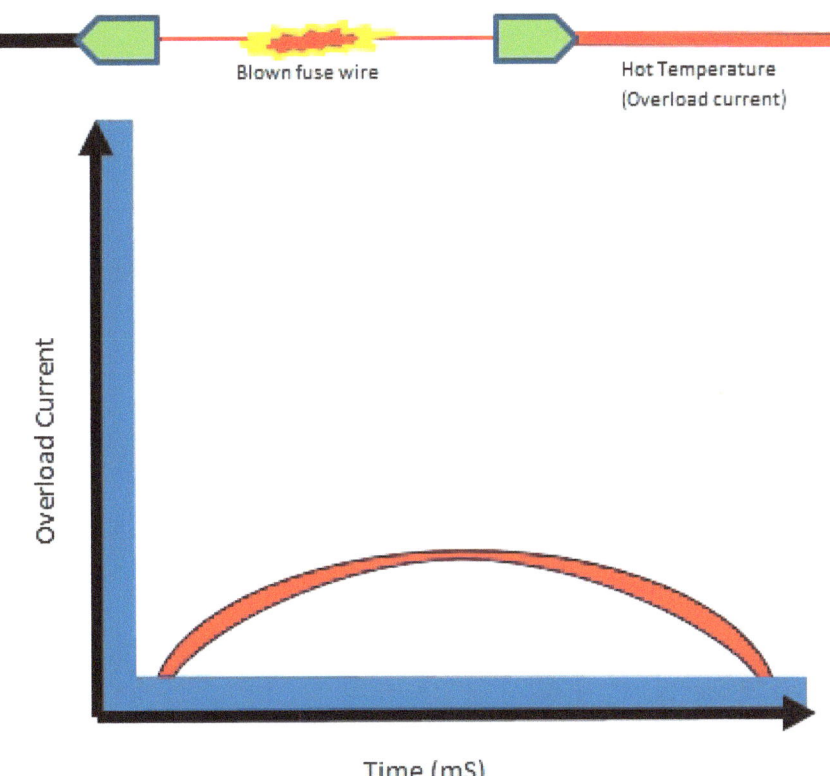

Figure 2: Overload Current

2.4 Overcurrent

A fault current is a much more violent event caused by either a phase to earth fault or phase to phase. If a mechanical digger accidentally strikes an underground cable or in a domestic premises a nail is put through a live cable this will cause an overcurrent fault. This is also commonly referred to as a short circuit or in BS7671 the Prospective Fault Current (PFC).

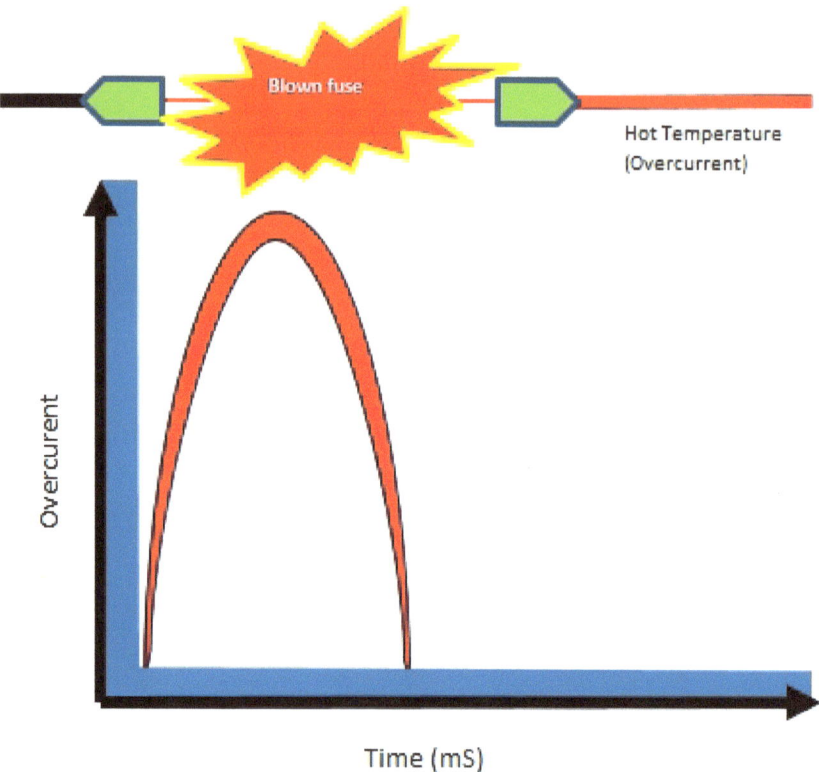

Figure 3: Fault current

From the diagram above the temperature increase is extremely rapid resulting in complete destruction of the fuse wire.

2.5 Earth Fault Protection

An earth fault can be either one of the following occurrences:

- Conducting material (e.g. metal enclosures, pipe work) becomes live when it isn't normally supposed to (indirect contact).

- The insulation surrounding live conductors (e.g. a buried power cable) degrades and allows leakage current to flow to the surrounding soil particularly in damp conditions (indirect contact).

- Earth leakage when motor and circuit insulating materials degrade slowly over a period of time.

- Or a human being or livestock inadvertently makes contact with a live exposed conductor (direct contact)

Although at first glance an earth fault can appear the same as the previously discussed overcurrent phase to earth fault, in this context we are discussing earth faults that are smaller leakage currents to earth from the live conductor, measured in milli-Amps (mA). Remember that the overcurrent or PFC event happens instantly with much higher fault currents than earth leakage detection values. Earth fault protection systems are set to at much lower detection values i.e. mA than overcurrent devices.

Domestic installations and other electrical installations will require high sensitivity Residual Current Devices (RCD) that are rated to detect and trip at earth leakages less than 30mA. This value is given due to the average human being experiencing ventricular fibrillation at approximately 16mA.

The following diagram describes a method of detection and protection for earth leakage faults.

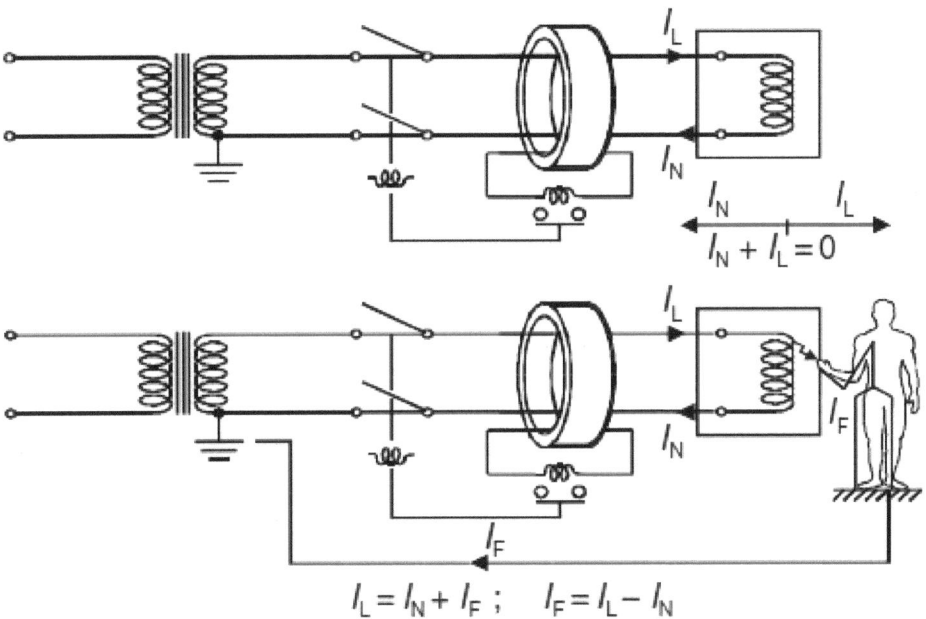

Figure 4: Earth leakage (courtesy of Power System Protection by Hewitson & Brown, 2004)

From the above figure, when there is no fault current, line and neutral currents are equal and the Current Transformer (CT) will not detect any current. E.g. "what goes in must come out" if current is leaking to earth above a certain value the CT will detect it due to the imbalance and thus trip. To put another way when the human touches the live part (not recommended!) the CT will detect current flow through the human and to earth (electricity will always find the easiest path to earth). As soon as the leakage current flowing reaches approximately 30mA the CT will see an imbalance and trip the circuit within 300mS. 30mA is the recommended maximum threshold for human beings.

Motor earth faults typically involve insulation degradation over a period of time as is the same for circuit cables. Unlike LV circuits outlined earlier that require 30mA earth leakage protection, motor circuits simply would not start with 30mA earth leakage and instead have their earth fault monitoring set at between 10 & 20% of full load current (FLC).

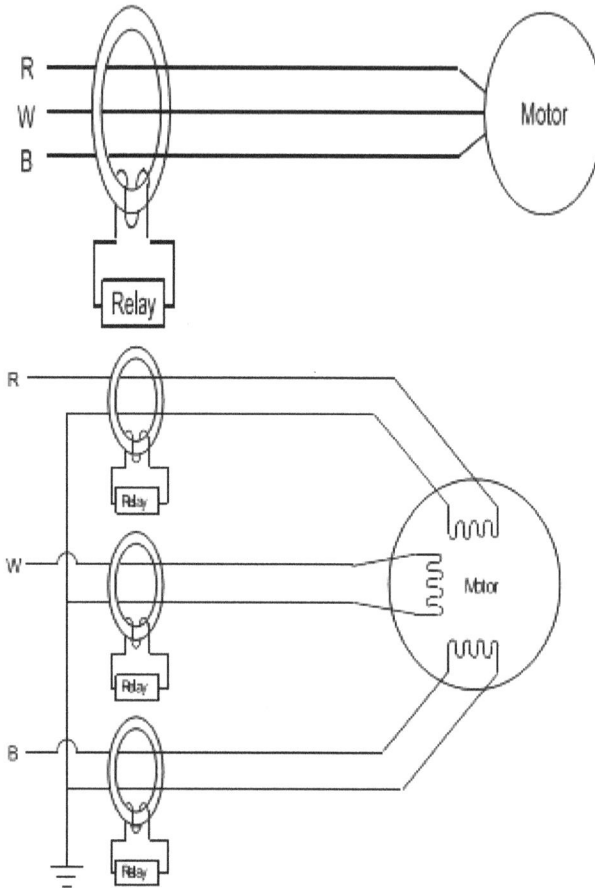

Figure 5: Motor earth fault protection

Motor circuits are monitored for earth leakage in a similar way to RCD's with a CT monitoring imbalances in the circuits but at a higher value.

To summarise earth faults:

- When the fault current flows through the earth return path it is called an "Earth Fault"

- For the purposes of this document overcurrent faults for switchgear and motors are instant and much higher fault current even though the fault current can more than likely flow to earth (phase to earth faults or more common than phase to phase).

- Earth fault monitoring and protection will monitor and interrupt much lower values of fault current than an overcurrent device

- Domestic installations and most other LV installations require earth fault protection in the form of suitable earth bonding and or RCD earth leakage protection at 30mA.

3.0 Components and Methods of Protection

3.1 Protection System

A protection system is designed to detect a fault condition and depending on the nature and type of fault to quickly disconnect the supply feed from the circuit. The protection system can be one device e.g. a fuse on its own or a combination of fuses, relays, CT's and circuit breakers depending upon the application and fault type.

This section will provide an indication of the various components for protection systems typically encountered during routine maintenance and fault finding. It will provide an overview of contactors for comparison purposes only. More concise guidance will be provided in "Electrical Switchgear Maintenance Part 2"

3.2 Characteristics

The purpose of a protection system is to protect against unplanned failures of supply whilst ensuring safety of personnel and avoiding damage to infrastructure. The basic requirements for any protection system are listed below.

3.2.1 Discrimination

For multiple circuit system the protection system must isolate the faulty circuit only.

3.2.2 Stability

The protection must remain stable up to the maximum allowable fault value and not cause "nuisance" tripping. An example of this during a motor start up the starting current can be 6x its normal running current or fluorescent lighting circuits that have initial surge currents when switched on.

3.2.3 Sensitivity

As well as being stable the protective device must operate at appropriate levels of fault value.

3.2.4 Reliability and Speed

It must operate in the event of a fault and within the required time. The speed is vital due to the increase in damage at the actual fault location. This is the fault energy = $I^2 * Rf * t$ (I^2 = fault current^2, Rf = resistance of fault, t = time seconds).

Correct designs, frequent maintenance, testing and inspection will help ensure correct speed of operation and reliability.

3.3 Fuses

Fuses are self-destructive and can be used as part of a protective system or simply to protect circuits that do not require specific circuit monitoring as highlighted earlier.

In HV motor circuits fuses are used in series with contactor circuits to provide overcurrent protection as contactors are not designed to break overcurrent faults, circuit breakers are. At LV levels fuses can also be used to back up circuit breakers to interrupt high level overcurrent that the circuit breaker cannot handle.

Different types of fuses are available such as BS88 etc. BS88 fuses are capable of providing large values of overcurrent protection as well as normal overload duties provided they are specified correctly for the circuit. In a domestic premises or similar a simple fuse wire will only interrupt low fault currents or a Miniature Circuit Breaker (MCB) will interrupt lower fault currents than the main incomer BS88 fuse. Hence sometimes the main incomer fuse of a house will blow before the MCB in the event of an overcurrent.

3.4 Circuit breakers

Circuit breakers are designed to withstand high fault currents and remove the fault current in rapid movement. For ease of reference we will discuss circuit breakers rated at 3.3kV to 11kV Circuit breakers of this voltage rating contain mostly mechanical parts with electrical components used for closing, tripping and monitoring. This is due to the high closing forces and speeds required at higher current ratings than lower rated circuit breakers. The main components of a circuit breaker are:

- Circuit and busbar spouts
- Charging springs for closing and opening
- Charging motor
- Limit switches, auxiliary contacts and status indicators
- Solenoid release shunts.
- Arc interrupters or arc interrupting medium

The charging springs (open and close) are used to fix the contacts in either position. They are also used to provide the required speed and force to close or open the contacts. The charging motor re-charges the springs in readiness for the next operation (close or open). The shunt release coil will release the closing or opening springs.

The volumetric size of circuit breakers can vary from approximately $1m^3$ to $3m^3$, figure 6 shows a 3.3kV Vacuum Circuit Breaker (VCB) that is approximately 2m height by 0.5m width, 0.7m depth and weighs 300kg.

Figure 6: 3.3kV Vacuum Circuit Breaker

Circuit breakers have moving and fixed contacts designed to open or close under fault conditions. During a fault the current is very high and opening against a high fault current will produce a large arc that must be extinguished very rapidly. This means that action of the fixed and moving contacts must be capable of:

- Very quick opening and closing
- Maintain a firm and contact when closed
- Having low resistance across the contacts when closed (usually 20 to 300 µΩ depending on the age of equipment).
- Break the fault current without explosion

You can see from figure 7 below the mechanical content required to make and break the circuit breaker contents. This is the closing and opening mechanism for a Siemens Sion 11kV VCB. The manuafacturer quotes a closing time of <75mS and an opening time of <65mS for this VCB.

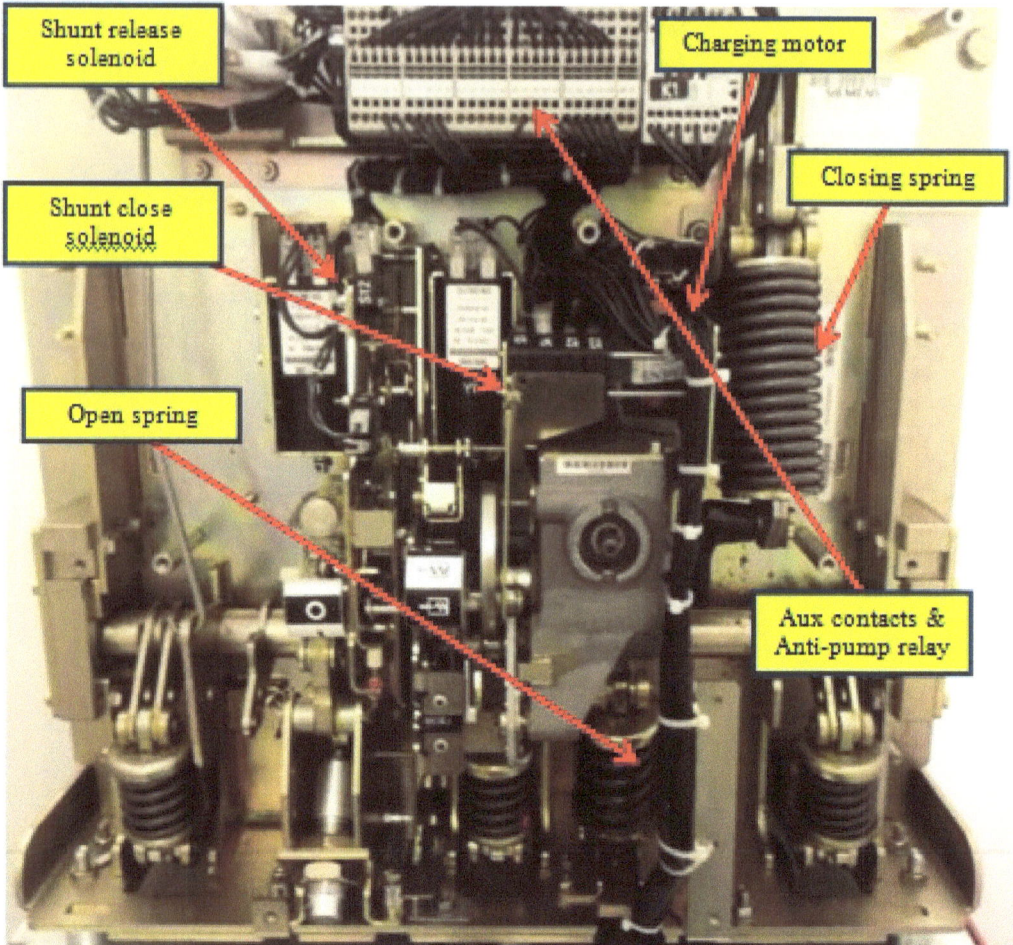

Figure 7: Internal of Siemens 11kV VCB

To extinguish the arc there are arc chutes to break the arc into smaller segments and dissipate the energy created safely. Also used to extinguish the arc are different types of insulating mediums surrounding the contacts such as air, oil, vacuum and Sf6.

Salient points regarding arc interruption methods are as follows:

(a) Oil circuit breakers

Oil circuit breakers although still in use today but are being phased out due to hazardous waste concerns with the oil and with the much more efficient vacuum and Sf6 circuit breakers.

For an oil circuit breaker the contacts are immersed in oil to extinguish the arc during opening and closing. Oil circuit breakers have good cooling properties but require more maintenance (frequent oil changes, oil testing) than other types.

(b) Air Circuit Breakers

Air Circuit Breakers (ACB) use air as the extinguishing medium with arc chutes to dissipate the arc. ACB's contain large fixed and moving contacts that require regular maintenance.

(c) Vacuum Circuit Breakers

Electricity cannot exist within a vacuum and therefore the contacts can be many times smaller than the equivalent rated ACB. There are less moving parts and no arc chutes. VCB's were slowly introduced in the 1960's and became more popular in the 1990's once the initial problems had been resolved. Out of all the circuit breaker types listed they are by far the most superior in terms of reliability, number of operations, life span etc.

(d) Sf6 circuit breakers

Sulphur-hexafluoride circuit breakers are normally used at voltages of 22kV and above. Sf6 is excellent at extinguishing contact arcs formed at higher voltages. Unfortunately it is an ozone depleting gas and additional precautions have to be taken when maintain Sf6 circuit breakers, for example the Sf6 is decanted into a sealed container and the same amount must go back into the circuit breaker. Further information is available on Sf6 circuit breakers from other publications and is not within the scope of this paper.

To summarise the different mediums available for circuit breakers:

Medium	Risks	Maintenance	Reliability
Oil	Fire, disposal	Oil changes	Low/good
Air	Hot air to surrounding area (if arc chutes fail)	Arcing contact wear	Good
Sf6	Environmental if seals fail	Minimal , lubrication of mechanisms/visual	Excellent
Vacuum	Loss of vacuum	As above	Excellent

Table 1: Circuit Breaker Medium Comparison

3.5 DC Trip Circuit

A key component of the protection system is its tripping supply. In the event of power loss a reliable source is needed to trip the circuit breaker. This is normally at 110VDC and provided by an uninterruptable power supply (UPS) consisting of a combination of batteries, chargers and micro generators. Some systems require that during a minor power dip (μ seconds) the motor circuit must immediately restart or not be allowed to restart depending on the application. This is the reason that mechanically latched circuit breakers and contactors utilise DC as the control circuit.

A DC source can also be used for critical back-up lubrication supplies if AC powered lubrication systems lose power.

3.6 Relays

The main types of fault protection relays referred to within this paper are overload, overcurrent and earth fault. There are of course many other types of relay such as under voltage, supply fail, inter-trip etc. for simplistic reasons only the three types of fault protection will be discussed throughout this paper.

Figure 8: P&B Golds Overload Relay

3.6.1 Overload

An overload relay will simply trip during excessive loading of a circuit. The overload relay graph in figure 8 shows that at 115% overload setting it will trip after 28 minutes. It will also allow certain starting currents without tripping instantaneously. For example during starting it will allow 6* the 115% starting current for 1 minute.

3.6.2 Overcurrent

An overcurrent relay will detect much higher currents than an overload relay and shouldn't trip during starting. They can be either instantaneous (approx. 300mS) or of the inverse (delayed but the higher the current the quicker the reaction) type depending on the application. Worked examples of both types are provided further on in this document. Overcurrent relays will be reviewed in more detail in part 2 of this series of switchgear maintenance papers. To summarise and overcurrent relay should only operate during high fault currents, be set below the calculated fault level (as close to the maximum start up current as possible) and not operate during start up.

3.6.3 Earth Fault

Earth fault relays can either be instantaneous of inverse, again depending on the application. As stated earlier they are usually set to allow motor operation without nuisance tripping during start up. A stabilising resistor is also installed to absorb motor start up current spikes.

3.7 Purpose of Current Transformers

CT's are used to transform high current values to lower ones for monitoring load conditions. One important safety point to note, if the secondary side is disconnected for any reason (e.g. to replace a faulty ammeter) secondary of the CT must be shorted out otherwise the CT will become unstable with primary current flowing and very high voltage will appear at the secondary terminals.

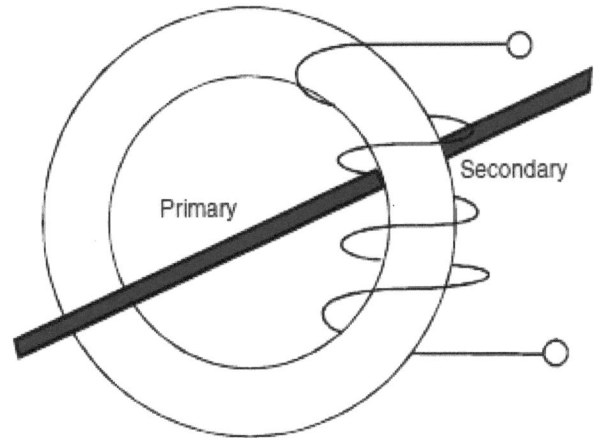

Figure 9: Bar primary CT

3.8 Protection Systems Recap and Generic Values

To summarise overcurrent protection monitors high value fault current or overcurrent (generally 8*FLC) for phase to phase and phase to earth. Earth fault protection will monitor lower value fault current at normally 20% of full load current for motors and mostly 30mA leakage current for circuits where human contact is possible. Motor circuit earth protection is not set at 30mA due to the transient currents that flow during start up and running of inductive loads. The motor would simply never get started. It is expected however that motor circuit terminals and all other live conductors are inaccessible to during normal operation, and if a fault does occur the structure is suitable bonded and the earth fault protection to operate almost instantly.

Primary injection tests will test the entire protection system including circuit breaker and protective relaying. Secondary injection will test the protection relays only and if let in circuit can also prove that the circuit breaker itself will trip.

	Contactor	Circuit Breaker
Fault break capacity	4.0kA	40kA
Current rating	400/630A	3000A
Contact gap 11kV	6.0mm	16.0mm
Contact force	10kg	80kg
No of operations	1-2.5M	10k

Table 2: Comparison of vacuum contactor and vacuum circuit breaker

Circuit breakers are capable of larger fault current breaking but lower operations than contactors. Contactors are typically used for motors that are continually switched on and off. Contactors are backed up with fuses for overcurrent protection. Operations staff can often be puzzled as to why the protection relay hasn't tripped the contactor in the event of a fault. This is because the contactors are not capable of switching overcurrent faults and the fuses are. The fuses usually have a striker pin that will activate the protection relay to prevent the contactor attempting a reclose (restart). Otherwise the motor might be restarted with an unbalanced supply e.g. two phases.

4.0 Circuit Breaker Maintenance

4.1 General

The following is a suggested routine and OEM guidelines should be referenced at all times. Circuit breakers (and contactors) above 300A will contain many moving parts such as; mechanical interlocks, levers, springs, limit switches and solenoid coils. Before performing internal maintenance:

- Safety – ensure; charging springs are fully discharged, racked out from its cubicle, busbar shutters locked shut (or the circuit breaker can be removed from the switch room itself).
- Double check springs are discharged by attempting mechanical close & trips

4.2 Visual/tactile

(a) Externally check insulated components, cast resin parts for cracks, cleanliness, damage and securely fastened

(b) At the interface between insulated and non-insulated check for evidence of partial discharge activity

(c) Loose wiring, harnesses snagging moving parts, overheating and corrosion

4.3 Mechanical

(a) Clean arc chutes (if fitted)

(b) For ACB's check clearances and timing of contacts. For example a 3 pole set of moving contacts should just touch the fixed contact within 0.015" of each other.

(c) Lubricate all moving parts with 20W50 or OEM recommended grease

(d) Strip down and clean cluster fingers as required replacing worn parts.

(e) Check and adjust all linkages, check for free movement and excessive play.

(f) Mechanically trip and close observing the operation of all moving parts including contact faces of ACB's.

(g) For VCB's consult the manufacturer's guidelines on checking vacuum bottle clearances.

(h) Cleaning using lint free rags on all insulated areas, no detergent to be used.

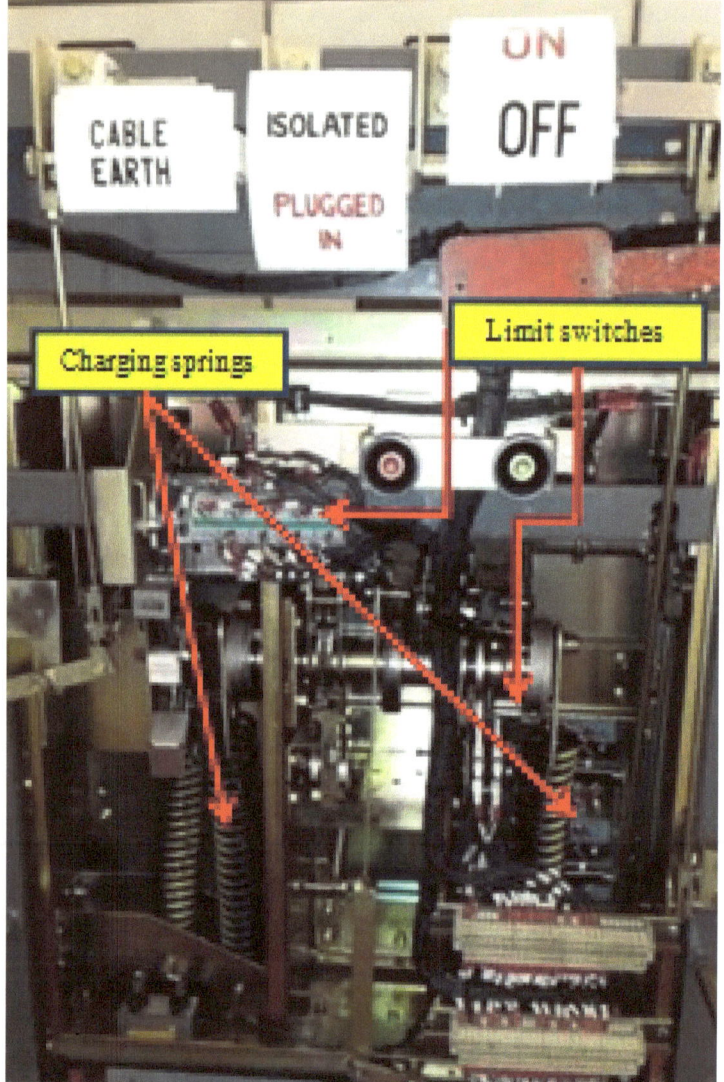

Figure 10: Sample 11kV Internal Charging Mechanism

To summarise the main focus of maintenance is cleanliness, lubrication and function checks. The frequency depends on use, e.g. if CB not used for 6 months it must be tripped and closed to confirm it still operates. High operational and high load use circuit breakers will also require reduced time intervals between maintenance schedules.

If the circuit breaker has acted for an overcurrent fault it must also be inspected along with its associated protective relay system.

4.4 Electrical Tests

There are three types of electrical tests are carried out on large circuit breakers; the continuity of joints in µΩ (commonly known as "ductor" tests), insulation resistance (IR) and HV AC test for earth leakage in mA. Polarization Index tests are also used for a more thorough test of insulation conditions.

(a) Ductor test

This low resistance test is used to prove good continuity of busbar joints, circuit breaker poles (across busbar spout to circuit spout) and earthing. Low resistance is classed as less than 1Ω. Frequent checking of the same joint or circuit breaker pole is used to monitor of the joints are degrading at all, e.g. an increase in resistance value can indicate joints becoming loose, corrosion between mating surfaces or over stressing caused by over loading. For circuit breaker maintenance the test is done before maintenance and at the end to prove how much circuit breaker contact faces and joints can degrade between maintenance periods.

Typical values for a Siemens 11kV VCB approximately 8 years old will vary from 11µΩ to 40µΩ at 1A test. The same test on a 40 year old 3.3kV ACB will vary from 120µΩ to 350µΩ prior to cleaning of the contact faces. All 3 poles must balance within 20% of each other.

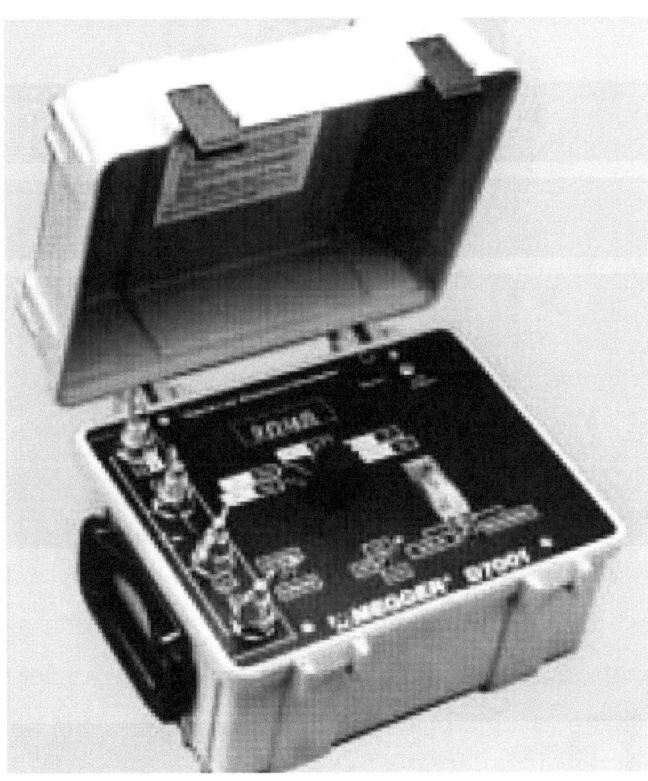

Figure 11: Megger D7001 Ductor tester

The test leads each have two probes, one set for injecting the current and another for measurement.

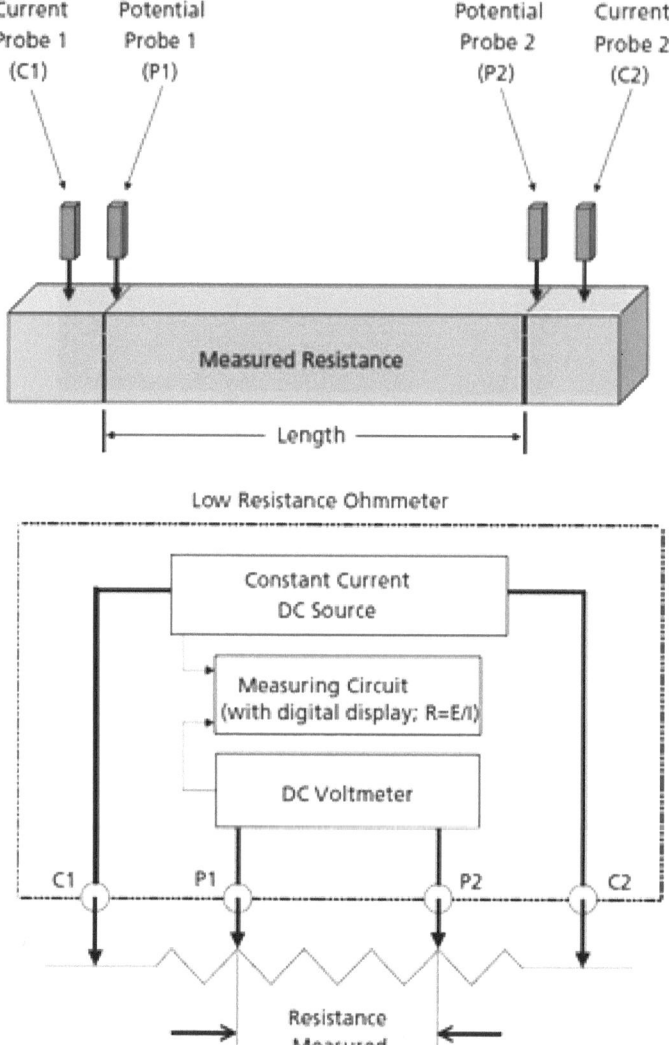

Figure 12: Ductor probe measurements

If higher than normal readings are found the joint must be stripped cleaned and retightened to the correct torque values. When testing across VCB's and one pole is considerably higher than the others the vacuum bottle itself might require changing if the connections leading up to the vacuum bottles have proven ok.

(b) Insulation Resistance

For IR readings the IEEE recommends the following guidance figures for HV equipment

- For minimum expected IR readings per phase approx. 9M Ω per kV rating of circuit breaker from phase to earth, e.g. 11kV should have min reading of 99M Ω (IEEE recommendation).
- General I/R Test voltages; >3kV rated = 2.5kV DC, 5kV – 11kV = 5kV DC.

(c) Polarisation Index Test

A polarisation index (PI) test will give more information e.g. if the insulation on the circuit breaker bushing has failed etc. A PI test is a ratio of two IR tests carried out for 1 minute and 10 minutes with the 10 minute test results being divided by the 1 minute test results. When a DC test voltage is applied, the current the flows is broken into 3 parts:

- Charging current – the current that charges the capacitance of the system which starts at maximum then reduces to negligible proportions in a short time
- Absorption current – current flowing within the insulation and will decay with time
- Conduction current – current that flows over the surface of the insulation

From the above the current flowing will decrease with time and the apparent IR will increasing. The IR increase will be very rapid at first as the charging current decays. If the insulation is undamaged, clean and dry it will take longer for a steady IR value to be reached. This is due to the "conduction current" being small compared to the "absorption current". Therefore if the insulation is in poor condition the conduction current will be high resulting in a steady IR value being reached quickly. From this it can be seen that when the ratio of apparent insulation resistances taken at 1 minute then 10 minutes will indicate the state of the insulation from the rate of rise of the apparent insulation. Typical expected values for good PI ratio:

- HV switchgear = 1.0 to 1.1
- HV motors = 2.0 to 3.0
- HV cables = 3.0 to 4.0

PI readings should always be compared with historical data if available.

If the readings are below the required standards then the equipment should be thoroughly cleaned (as described earlier), dried out and re-tested.

(d) HV test

The HV test is also referred to as an "overvoltage" test using an AC HV test set to test the integrity of insulation and vacuum bottles. Normally done after ductor and IR tests

For AC overvoltage tests the following guidance figures for HV equipment at one minute:

- 2.5 – 5kV circuit breakers = 12.25kV AC test
- >6kV circuit breakers = 19kV AC test
- 11 – 14kV circuit breakers = 27.0kV AC test.

General I/R Test voltages; >3kV rated = 2.5kV DC, 5kV – 11kV = 5kV DC

The following table lists different test parameters for each test for an 11kV Vacuum Circuit Breaker (VCB) requiring a 4 year overhaul. Note these values are not for new, but already in service circuit breakers. The order of tests should follow the sequence shown in the table.

Test	Duration	Comments	Expected Results
Ductor test	Until readings settle	Using ductor, VCB closed (after manually opening and closing 3 times)	Readings across all 3 phases within 20%of each other
IR to earth	1 minute	IR Tester at 5kV, VCB closed, 1 up 2 down to earth	99M Ω
HV AC Test (to earth)	1 minute	HV Tester at 27kV, VCB closed, 1 up 2 down to earth	0.6mA (from actual records)
IR to earth	1 minute	IR Tester at 5kV, VCB closed, 1 up 2 down to earth	99M Ω
IR across interrupter	1 minute	IR Tester at 5kV, VCB open, across each pole (e.g. L1 to L1)	99M Ω
HV AC Test (across interrupter)	1 minute	HV Tester at 27kV, VCB open, across each pole (e.g. L1 to L1)	0.4mA (from actual records)
IR across interrupter	1 minute	IR Tester at 5kV, VCB open, across each pole (e.g. L1 to L1)	99M Ω

Table 3: Circuit Breaker Electrical Tests

5.0 Protective Relay System Maintenance

5.1 Relay Mechanical Inspection

The mechanical checks apply to electromechanical relays only and include the following checks:

- Wipe action, the action of the moving contact "wiping" against the fixed contact and correct alignment

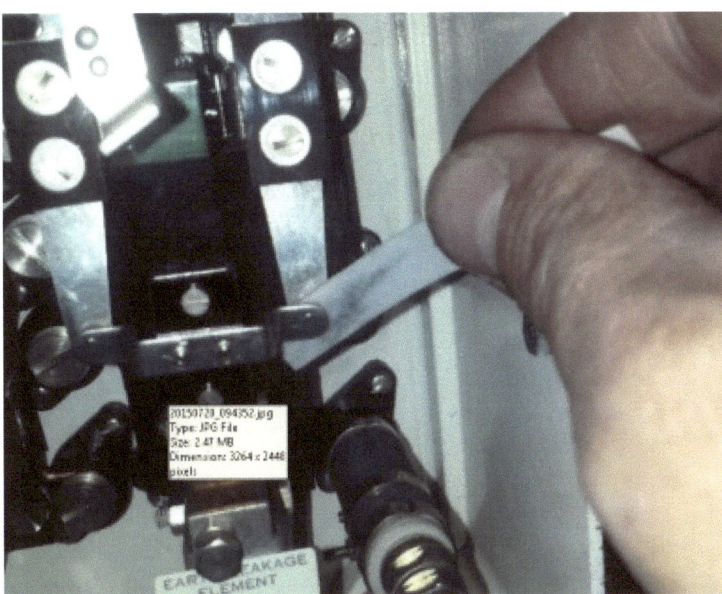

Figure 13: Contact cleaning strips

- Cleaning with contact strips and or fibre pens
- Terminal tightness

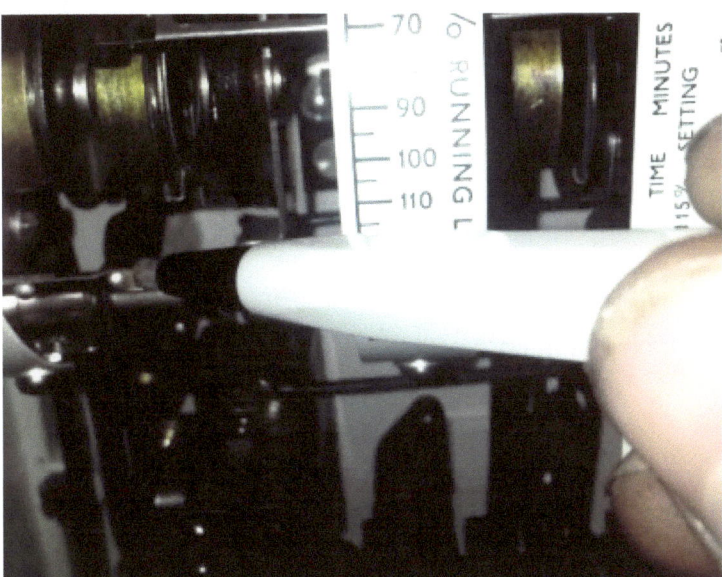

Figure 14: Fibre pen for cleaning contacts

5.2 Secondary Injection of Relays

Secondary injection tests are used to prove that the protective relay is operating within limits and is also a method of calibrating the relay. It is usually done in conjunction with relay mechanical checks. Typical tests include:

- Proving the auxiliary flags will operate at <66Vdc (electromechanical types only) by slowly injecting DC voltage and simultaneously holding one of the fault contacts
- Injecting earth fault current directly onto the E/F secondary terminals, for example a 5A relay set at 20% E/F should trip at 1.0A +/- 5%
- Injecting overcurrent directly onto the R phase secondary terminals, for example a 5A relay set at 8* should trip at 40A +/- 5%
- Proving that the relay will trip the circuit breaker

Detailed examples are provided at the end of this paper.

5.3 Primary Injection Tests

Primary injection is the injecting of full load AC current using a specialised test set through the primary circuit and proves correct functionality the entire protection circuit. It is difficult (and most cases) impossible to duplicate the overcurrent values through primary injection due to the high currents required. For example a 150A FLC motor set at 8* for overcurrent would require a test set to produce 1200A! Primary injection is very good at proving the sensitivity and stability of CT's including the secondary circuits. Primary injection of a protective system is performed when:

- Initial installation or replacement of CT's
- Secondary circuit wiring changes
- If the CT's are suspected as being faulty (e.g. intermittent trips during start up)

(a) CT Polarity test

If new CT's are fitted it is necessary to check for correct polarity when installed, failure to do this will result in tripping of the protection system due to incorrect polarity.

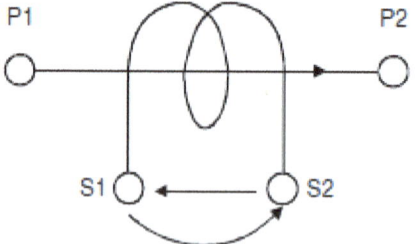

Figure 15: CT Polarity Configuration

Using a 6v battery and moving coil AVO make connections as follows:

- + from battery to P1 side of conductor flowing into the CT
- - from battery to P2 side of conductor flowing out of the CT
- + AVO to S1 on CT
- - AVO to S2 on CT

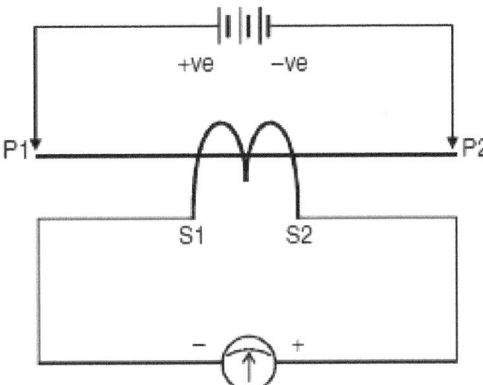

Figure 16: CT Polarity test

Set AVO to lowest DC current value and momentarily connect one of the battery leads whilst observing a slight flick on the AVO in the + direction.

(b) Primary injection

Primary injection is used to test stability, e.g. no spill current short out all 3 phases and inject from L1 + to L2 -, L1 + to L3 -, L2+ to L3 –. There is also a test for sensitivity by connecting heavy current from L1+ to L1-, L2+ to L2- and L3+ to L3-.

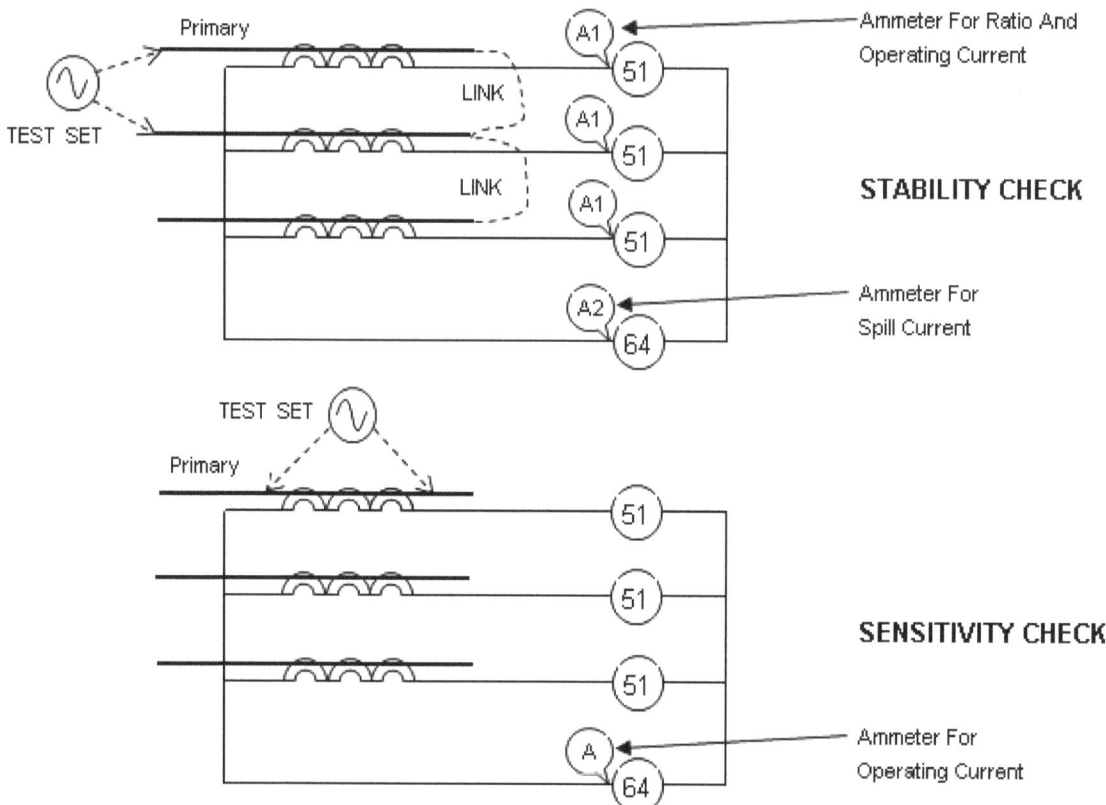

Figure 17: Primary Injection Stability and Sensitivity Checks

A calibrated clip on ammeter is usually applied to the secondary of the circuit under primary injection to monitor spill current and sensitivity. The protection relays are left in service with alarms and values noted during the course of testing as required. Once again detailed examples are provided at the end of this paper.

(c) Primary injection summary

- The stability test is used to prove that the system remains stable during normal running and will not trip on earth fault i.e. there is no spill current.

- The sensitivity test is to confirm that the E/F operates within required limits

6.0 Real Life Scenarios (Protection Testing, Commissioning and Maintenance)

6.1 English Electric 3.3kV Maintenance

(a) Preface

This type of circuit breaker is an English Electric class 'E3' Air Circuit Breaker manufactured in 1970's and still in use today (although many have been converted to VCB's). It covers the various current ratings up to 1,600A and both solenoid and spring closing. The following is a sample overview of the maintenance required for a 4 year overhaul. The list is not exhaustive and is based on current practices at the time of writing at one particular location. The example is provided to give the non-experienced reader a feel for the work required to maintain an Air Circuit Breaker (ACB)

Figure 18: English Electric Air Circuit Breaker

(b) Equipment

- 2.5kV Insulation Resistance Tester
- Ductor tester
- Hand tools

(c) Method: Initial Checks & Cleaning

- Clean down and inspect the ACB as strip down is progressed.
- Prior to starting work, inspect floor of cubicle (if access available) for signs of debris or gearbox oil that may have fallen from ACB.
- Extend the Circuit (top) and Busbar (Bottom) Isoselectors in preparation for carrying out the IR tests and tilting back the Arc Chutes.
- With the ACB open, I.R. across the ACB at 2.5kV on each phase.
- Close the ACB and I.R. to earth and between phases at 2.5kV as follows:
- Earth the Yellow and Blue Phases. I.R. the Red phase to earth.
- Earth the Red and Blue Phases. I.R. the Yellow phase to earth.
- Earth the Red and Yellow Phases. I.R. the Blue phase to earth.
- Ductor ($\mu\Omega$) across each phase in turn on 10A range and record results.
- Remove the top front cover plate to give access to the Black plastic channel insulating screens.
- Remove the screens and release arc chute continuity links so that they can be pivoted out of engagement. Hinge the Arc Chutes backwards so that they rest on the circuit isoselectors.
- Unscrew and remove the Isoselector and truck locking bar operating knobs by first removing the Split pins. The lower front cover should then be removed.
- Check Isoselector operation including Circuit, Busbar and Earth selector mechanism and Associated Mechanical Interlocks. Also check Semaphore operation.
- Visually examine each arc chute and report any damage found.
- Note: If the arc chutes have been removed from the ACB they should be stored within the ACB cubicle with the heater switched on.

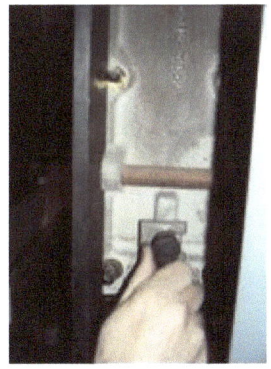

Figure 19: E3 Arc Chute Removal & Contacts Access

- Remove the insulated side plates to give good access to the main contact assembly.
- Unclip and remove the lower Black plastic channel from each phase.
- Remove the screws at the rear of the ACB securing the side plates to the carriage and the insulated tie bar nuts. This releases the rear split barrier and the Aluminium air cylinder.
- Remove one of the two 2BA screws from the rear white barrier holding the two sides of the inner side plates together. The inner side and lining plates can then be removed by gently pulling them away from the insulated pegs

Figure 20: E3 Air Cylinder Piston

- Wipe out and inspect the air cylinder bore checking for signs of scoring and excessive wear. Also inspect the felt washer to ensure that it has not compressed to the extent that it has become ineffective. This can be checked by fitting the cylinder over the piston and operating, noting the airflow from the nozzles. Replace the felt washer if required.

- Inspect the insulated pegs in the arc runners and replace if damaged or broken.

- Clean ACB and visually inspect. Check in particular the following:
    o Ensure all fasteners are secure and split pins are correctly fitted.
    o Check that each of the push rods are not free to rotate on its pivot pin (as the internal screwed stud can become loose).
    o Check that the fixed arc runners and arc chute continuity connectors are tight.
    o Ensure that the air cylinder pistons are free on their pivots.
    o Visually inspect insulated bushings and push rods for chips, cracks and discolouration.
    o Inspect conductors for signs of overheating etc.

Clean auxiliary contacts and mouldings, checking connections for tightness and ensure that there are no signs of overheating, corrosion or Verdigris. Each of these switches can be checked and cleaned by removing the operating linkage from the switch shaft after first marking its position. This will allow the switch to be rotated through 360 degrees giving access to all the moving contacts.

(d) Method: Main Contact Assembly

- Examine for signs of excessive arcing, burning or misalignment of contacts.

- Remove any copper 'globules' from contacts and arc runners. Clean contact surfaces with a non-aggressive abrasive cleaner. If a plastic scouring pad is used the surface could retain a plastic film that could increase the contact resistance.

- On Circuit Breakers rated over 600A, bridge contacts are fitted to the moving arm as shown below. These and their fixed mating contacts have Silver contact pads that should be gently cleaned. There should be little or no signs of arcing if the circuit breaker contacts are set up correctly.

- Examine the flexible connections for signs of overheating, crimping or tearing of the laminations. Damaged flexibles should be replaced, as they cannot be reliably repaired. The connection screws should be checked if there is any discolouration at the edges.

- Ensure that the springs on the moving arm and the bridge contact are secure and in good condition.

- Inspect aux switches and wiring. Clean and lubricate as required. Also inspect, clean and lightly lubricate "jumper" contacts.

(e) Method: Contacts Setting

Slow close the circuit breaker until the moving contacts just touch the fixed contacts. They should all touch simultaneously within a tolerance of 0.4 mm (0.015 inches).

With the moving contacts just touching the fixed contacts the dimension from the front edge of the moving contact arm to the front face of the upper fixed contact plate should be between 24.0 mm (15/16 inches) and 25.4 mm (1.0 inch)

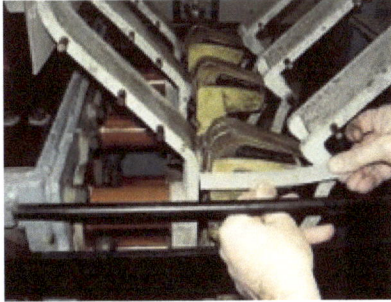

Figure 21: E3 Main Contact Setting

With the circuit breaker closed and latched the dimension in figure 21 should reduce to 12.7 mm (1/2 inch) on circuit breakers of 600 Amp rating and 14.3 mm (9/16 inch) on circuit breakers of a greater rating employing a bridging contact arrangement.

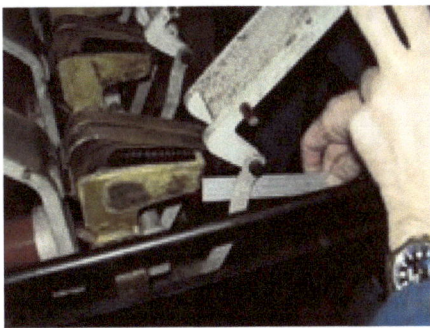

Figure 22: Underside Measurement

The measurement from the underside of the square interference nuts on the shackles to the bosses on the moving arc should be 14.3 mm (9/16 inch) on circuit breakers up to 600 amp rating and 12.7 mm (1/2 inch) on those of a greater rating employing the bridge contact.

Also, with the circuit breaker closed and latched, the dimension from the underside of the nuts securing the bridge contacts to the face of the moving arm should be 1.56 mm (1/16 inch).

If the above dimensions prove to be out of tolerance the contacts shall be adjusted as follows:

- Fine adjustment of the moving contacts can be carried out by rotating the nuts on the shackles.

- The nuts are chamfered to reduce stresses on the shackle studs and it is important to ensure that the chamfer is always horizontal.

- Adjustment of the contact arm is by means of shims beneath the insulation of the Push Rods. The nut beneath the Push Rod must be removed and the arm pulled free of the operating mechanism. The necessary thickness shim is then fitted over the stud and the Push Rod re-secured.

The above contact setting procedure must be repeated until the dimensions are within tolerance

(f) Method: Isoselector Contacts

The Isoselector contacts are spring loaded and designed to maintain a resistance free contact from the circuit breaker to the switchboard spouts.

Figure 23: E3 Isolector

Each of the six Isoselectors shall be stripped, cleaned and inspected for damage with components replaced as required.

 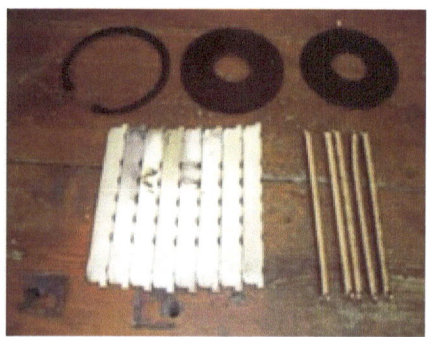

Figure 24: E3 Isoselector Assembly & Cluster Parts

Upon re-assembly minimal contact grease shall be applied. Too much grease will increase the resistance and obstruct up the moving parts.

(g) Method: Lubrication Details

POINTS OF LUBRICATION	RECOMMENDED LUBRICANT
Motor Charging Gearbox Oil	Shell Macoma 320 or Esso Spartan EP 320
Auxiliary and Control Switches Isolating Contacts Secondary Isolating Contacts Earthing Contacts	Pure White Vaseline or Best quality contact grease such as Electrolube CG53A. Colloidal Graphite in Oil, Arma 364

Table 4: Lubrication details E3 ACB

(h) Method: Final Checks

After the circuit breaker is re-assembled and cleaned:

- Ductor and IR the ACB comparing with the Pre-Cleaning values.
- Visual check of the circuit breaker, its cubicle ensuring all covers have been replaced, no loose fasteners, tools and materials cleared.

6.2 English Electric Magnetic Disc Type CDG 31

(a) Preface

The following maintenance procedure is for an English Electric (CDG) induction disc type for overcurrent protection.

(b) Connection details

For the CDG O/C relay terminals are dependent on the CT secondary wiring identifications used. Generally C12 = R, C32 = Y and C52 = B phases with the common being C73. The schematic diagram should be consulted for clarification.

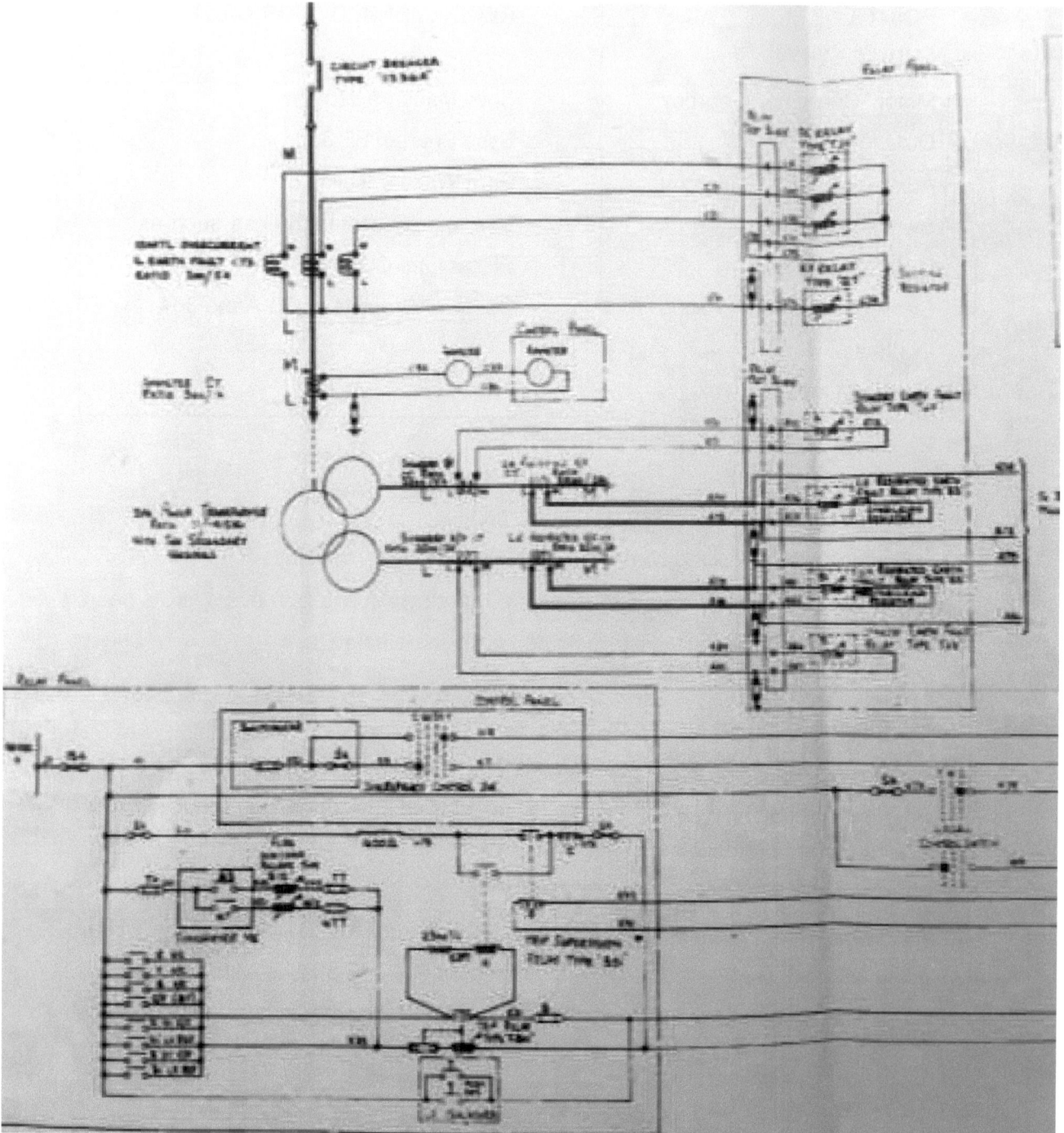

Figure 25: Schematic diagram of English Electric Protection

6.2.1 Procedure

Remove all fuses prior to commencing tests (a) to (d)

(a) "T" or "0" check and flagging

This check confirms mechanically that the contacts will make when the disc is fully turned and that the relay flag will drop (some flags will require the voltage to be present).

Set time multiplier to 0 or T and gently rotate until contacts make and observe the flag dropping. Also inspect the disc and contact faces for signs of pitting or damage. Clean with a soft brush or a feather is very effective.

(b) Disc return time

Again mechanical check to prove the disc returns when fully turned within a specified time. Proves that the spring is functional and the disc is not hindered. As above but time the return of the disc, usually around 12 seconds.

(c) Creep test

With normal running current applied the disc shouldn't turn.

Set the time multiplier to 1.0 and current plug to 5A. With 5A injected the disc shouldn't turn

(d) Energised reset time

Slowly inject 4.5A, at this current the disc shouldn't move. Rotate until the contacts make and time the disc return time.

(e) Overcurrent

To prove the disc operates at the specified load and within the required time. With the settings still as per above inject 10A and record the time taken to make the contacts. It should be approximately 10seconds.

(f) Trip test

To prove the protection relay will trip the circuit breaker replace all fuses and put the circuit breaker on "jumpers" (i.e. the circuit breaker auxiliary and control circuits connected not main power). Locally close the circuit breaker and carry out test (e).

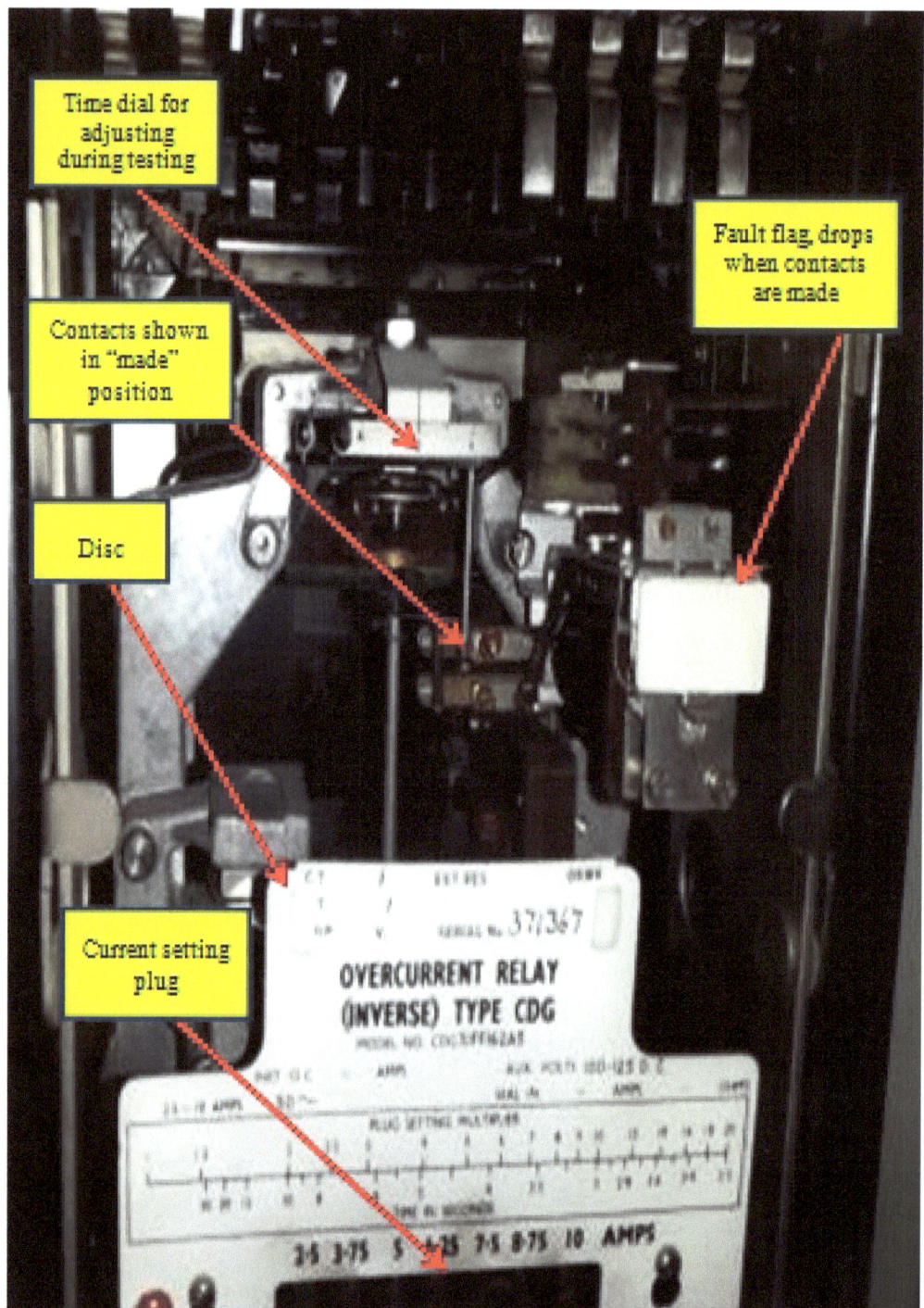

Figure 26: Reyrolle Inverse Overcurrent Relay

6.3 P&B Golds Mn4 Electromechanical Motor Protection

(a) Preface

This type of relay was manufactured from the 1960's to the 1990's and are still widely used today.

The Mn4 model has overload, instantaneous overcurrent, unbalance and instantaneous earth fault protection. It utilises bi-metal strips for monitoring current as opposed to the type from the previous section. One disadvantage with this type as compared to disc is that you have to wait for the bi-metallic strips to cool down and so they take longer to test. The disc type simply return to 0 after being injected.

Figure 27 shows a typical Mn4 relay manufactured by P&B Golds

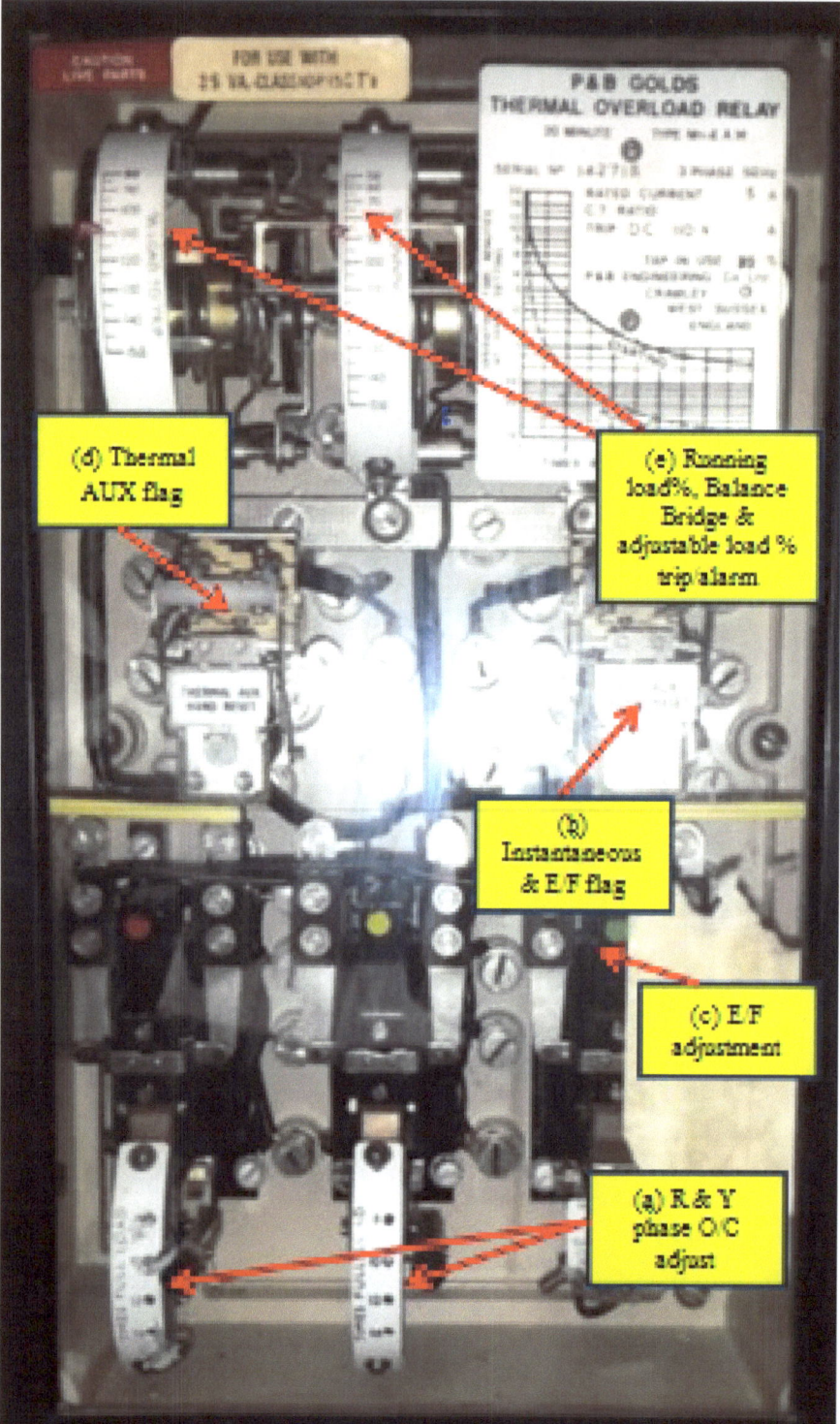

Figure 27: P&B Golds Motor Protection Relay

The overcurrent is monitored in the red and yellow phases only. Overload and earth fault across all three phases. Care must be taken when simulating overcurrent injection as the balance bridge can be become twisted due to only injecting one phase at a time. To overcome this, the current is injected and reduced very quickly so as not to damage the delicate instrumentation.

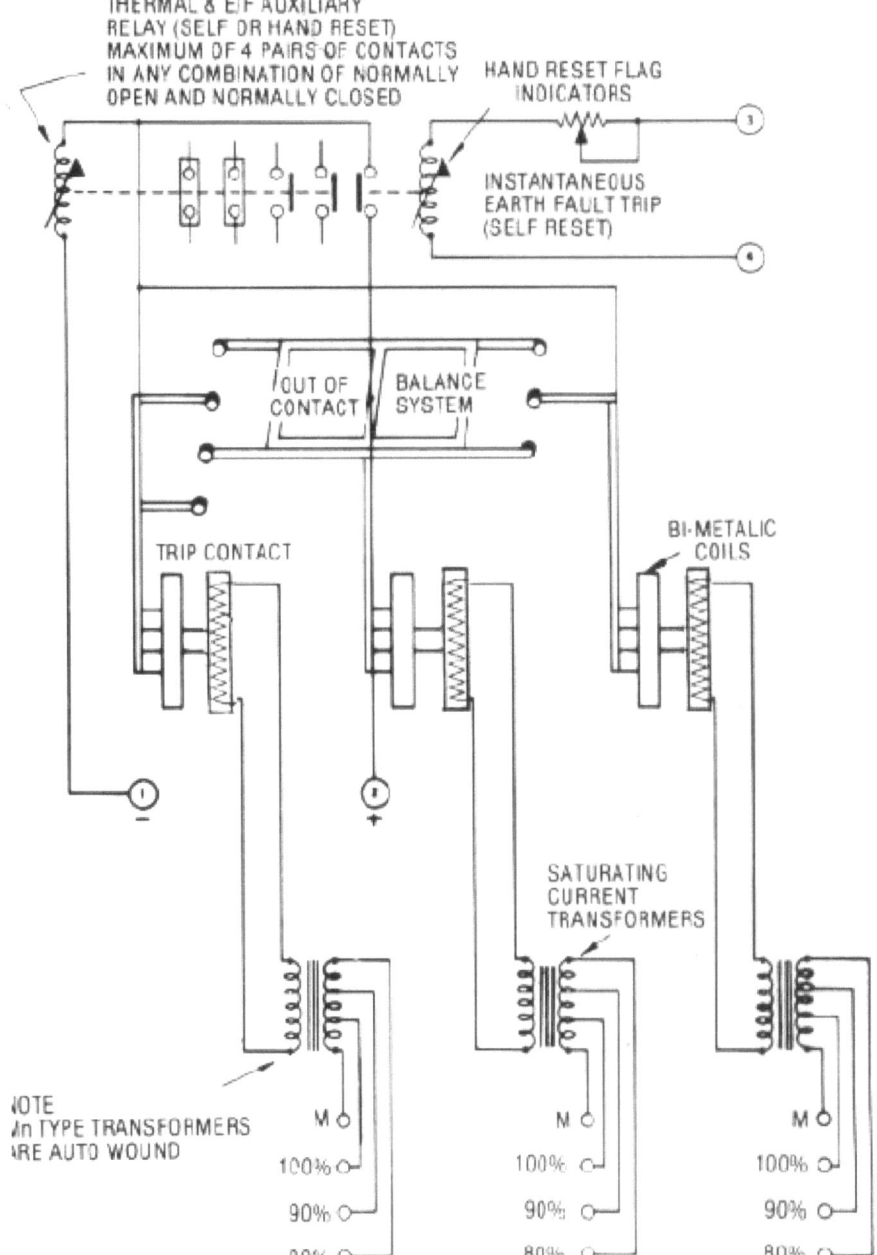

Figure 28: P&B Golds Schematic diagram

The schematic drawing above shows the internal wiring with out of balance contacts, overcurrent, overload and earth fault. The diagram below shows the actual connection details from CT to the relay on the motor circuit

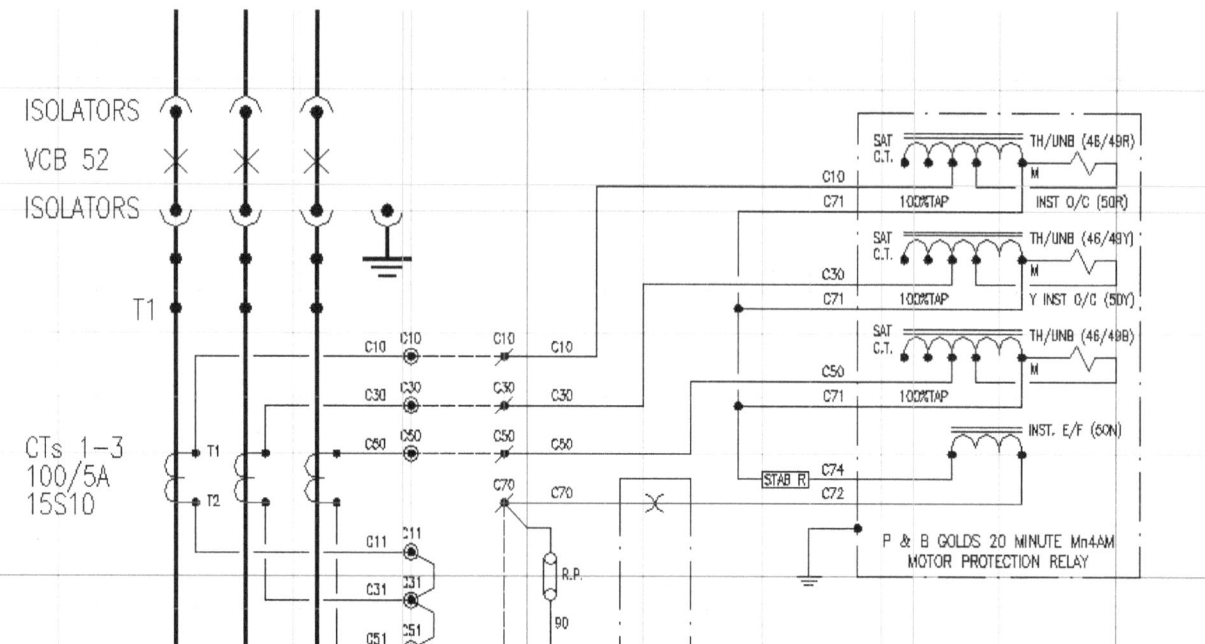

Figure 29: Schematic for Secondary Injection P&B Mn4

The following example is to maintain and test a protection system for a motor.

(b) Specific requirements

Ensure all fuses are removed and the circuit breaker associated with the protection relay is removed from the cubicle.

(c) Connection details

For each test connect to the relay terminals at the rear. Note there is no need to disconnect the outgoing field wiring in these tests:

Test	Connections (from the schematic)	Comments
(i) Overcurrent test	R = C10+, C71- Y = C30+, C71-	Overcurrent monitored in R&Y phases only
(ii) Earth fault	C74+, C72-	
(iii) Instantaneous relay (voltage) ,<66VDC	1-, 2+	R then Y O/C contacts closed
(iv) E/F relay (voltage) <66VDC	1-, 2+	E/F contacts closed
(v) Thermal relay (voltage) <66VDC	1-, 2+	T/C overload contacts closed

Table 5: Connection details

(d) Procedure

Note the settings; O/C, T/C, %Tap, 5A rated relay, 20% E/F rating.

Cleaning & mechanical checks:

- With the fuses removed, clean the contacts using a fibre stick and contact cleaning strips.
- Check each contact face and ensure the "wipe "action of moving contact to fixed contact is good
- Do not blow on the relay, remove dust with a soft brush

Instantaneous overcurrent tests

- Make connections as per (i) in the table above, drop the O/C to 8* and quickly inject 8* 5A= 40A. Do not leave on for too long as the contacts will distort. The R phase contact should make approximately 40A. If it is too low or too high there is a small adjusting screw to adjust the contact gap.
- Before injecting the Y phase wait until the relay has cooled down and measured value on the relay returned to zero or move onto another test

Instantaneous earth fault test

- Make connections as per (ii) in the table above, slowly inject up to 20% of 5A = 1A. Do not leave on for too long as the contacts will distort. The E/F phase contact should make approximately 1A. If it is too low or too high there is a small adjusting screw to adjust the contact gap.

Instantaneous E/F voltage relay:

- Make connections as per (iv)
- Slowly inject DC voltage,
- Should drop the instantaneous flag before 66VDC, record value

Instantaneous overcurrent relay voltage:

- Make connections as per (v)
- Slowly inject DC voltage
- Should drop the instantaneous flag before 66VDC, record value
- Instantaneous thermal relay voltage
- Make connections as per (iii)
- Slowly inject DC voltage
- Should drop the instantaneous flag before 66VDC, record value
- Tripping tests (to ensure the circuit breaker will trip via the protection)

Remove test leads, reset flags and replace fuses

With the circuit breaker auxiliary contacts made close the circuit breaker locally

Preform and E/F injection test and a phase O/C test. The circuit breaker should trip instantaneously.

6.4 MPR 3000 Electronic Protection Relay Commissioning

(a) Preface

When commissioning a new protection relay there are various parameters to be entered that are grouped into control settings, protection settings and system settings. Each group is further subdivided into other parameter sub-groups. For the purposes of this example sub groups "motor settings" (found under system settings) and protection settings will be outlined. The protection settings are concerned with how each protection parameter reacts and it is possible to disable parameters for the purposes of testing. The motor settings contain manually entered numerical values such as full load current (FLC) and CT primary etc. A detailed parameter list is normally provided by the designer or decided between the commissioning engineer and the client.

A new protection relay was required to replace an obsolete item with the following parameters requested by the client; CT ratio of 900:5, motor Full Load Current (FLC) 850A, t6X = 15 seconds, O/C = 300% for 10s, E/F set at 20%, High Set (HS) O/C at 1000%.

The secondary injection test was carried out on a mock up test bench used for training and setting up of relays. The CT's are individually connected for each phase with a common connection (C70 in this case) back to the relay. C70 is used as a residual connection from the 3 phase CT's as opposed to the other method using a core balance CT.

The various data sets are entered into the relay and then secondary injection tests carried out to prove that the relay will work at the specified parameters

This guide should be read in conjunction with MPR3000 technical manual available for download from P&B.

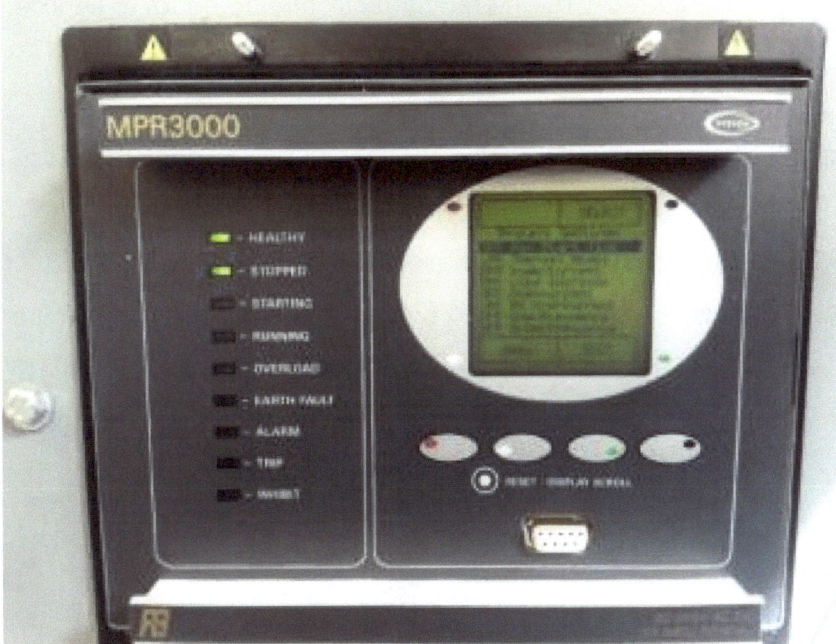

Figure 30: MPR3000 Relay

(b) Specific testing requirements

MPR3000 Relay Pre-requisites to enable testing:

- Pass word is ABAA [short out terminals 29/30 for MPR2000]

Adjust following parameters under the following menu-SETTINGS MENU →PROTECTION SETTINGS →

- Set "COOL TIME FACTOR" to lowest value (Thermal Model), Low Set Overcurrent (Overcurrent), U/V trip (Undervoltage) and max start time (Maximum start time)

- MPR3000 relays need to see a "simulated start, this achieved by quickly injecting 1.05* the FLC then returning back to required test value. It will then continue to see a running signal unless the current drops below 10% of FLC.

- Change "COOL TIME FACTOR" to lowest setting to speed up cool down of thermal curve.

To change a setting or enter new data the password ABAA will be requested prior to any change within that sub-menu.

To access settings for MOTOR data:

- SETTINGS MENU – SYSTEM SETTINGS – MOTOR SETTINGS, note the CT Primary value is greater than FLC

To access settings for PROTECTION data:

- SETTINGS MENU – PROTECTION SETTINGS – scroll to each relevant parameter

Ensure the parameters are correctly re-entered correctly prior to return to service.

(c) Connection Details

Using test block MMLB01connect THREE PHASE secondary test set as follows:

- Fixed 110VDC + to (2), 110VDC – to (1)

- Variable AC current source to RED phase CT secondary + to (21) and – to (22). For YELLOW phase + to (23) and – to (24). For BLUE phase + to (25) and – to (26). For E/F use + to (28) and – to (27). Variable setting to be 0 to 30A range.

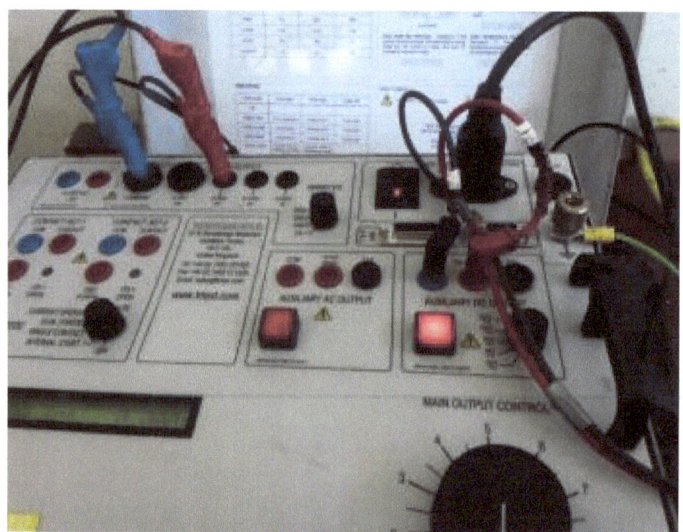

Figure 31: T&R 100ADM Mk3 Injection Test Set

Figure 32: MML B01 Test Block

Using test block MMLB01connect SINGLE PHASE secondary test set as follows:

- Fixed 110VDC + to (2), 110VDC – to (1)

Variable AC current source input + to (21) and return current source – to (23). Series link out (22) to (24), (23) to (26).

For E/F use + to (28) and – to (27). Variable setting to

Note, if the test block is not available use, disconnect each CT secondary at rear and short out the secondary wires from each of the CT's. CT wiring is usually identified as; red phase is C10, yellow C30, blue C50 and ground to C70. Specific wiring should be confirmed using a schematic drawing shown in figure

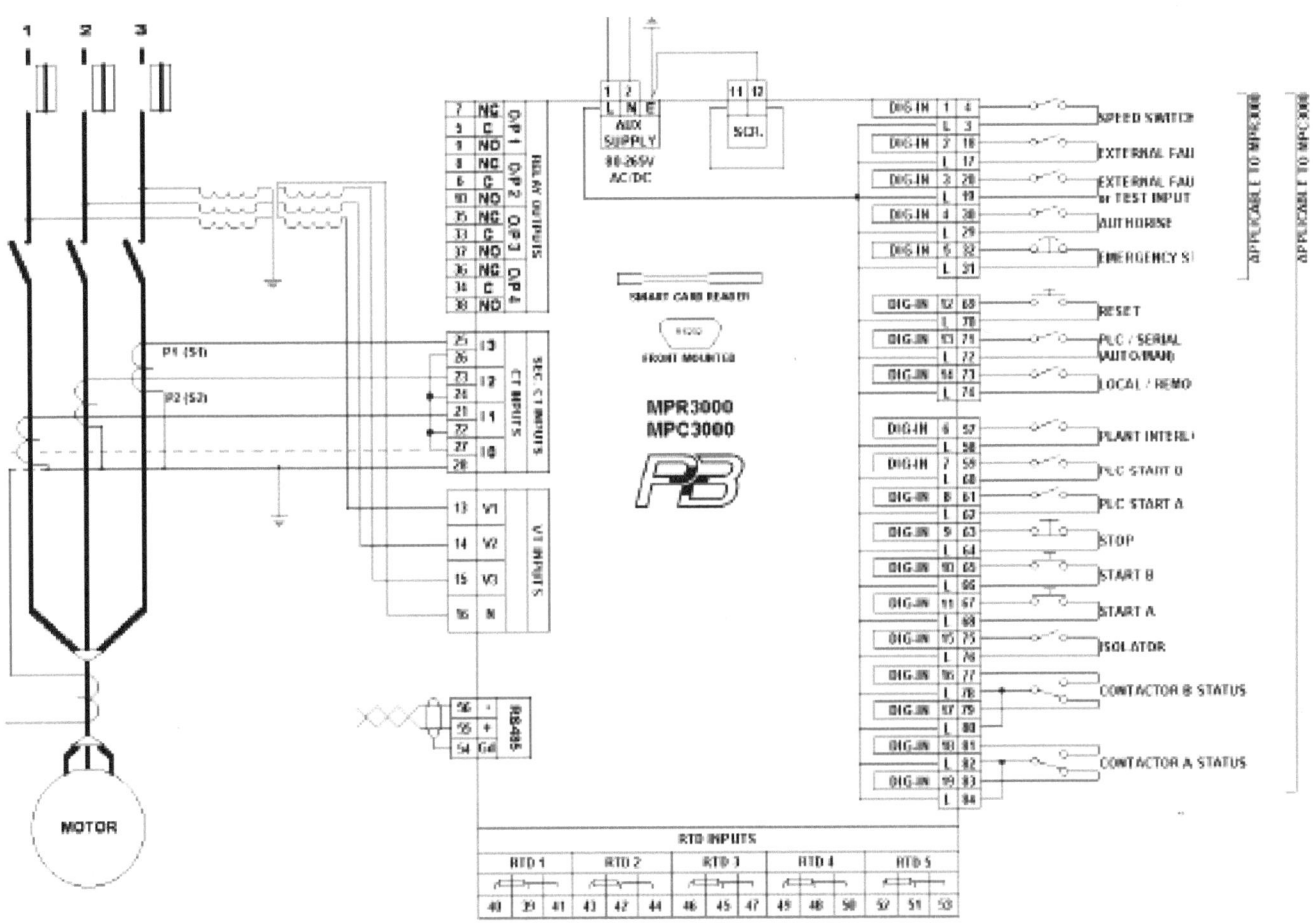

Figure 33: MPR3000 Schematic Diagram

(d) Procedure for MPR3000 Secondary Injection Testing

Enter data as required from previous sections and connect test leads.

Note that most electronic relays will calculate the % load over 3 phases, if <u>using a single phase injection set</u> the displayed current value will be 3 times higher.

- In motor protection = Parameter FLC at 850A, 5A relay, 900:5 CT
- Injection current of 4.4A = load value of 100%

(a) Overload tests (to prove the relay acts within the thermal curve T/C) with hot cold curve at temporarily set at 50% (accessed via motor protection settings):

- Find the "pick up" value of relay T/C whereby the load is 105%. Note the secondary injected value.

At 2* pick up value the relay should trip within approx. 14 seconds if t6x is set at 1second (value taken from appendix 5 of MPR3000 manual). However in this example t6x is 15 seconds. Therefore at 2* from app 5 = 6.046, multiply this figure by 15 seconds to equal 90.69 seconds. 2* 14.6A = 29.2A to inject. Repeat for each phase and record values. Note that even though the relay is electronic it mimics a mechanical bi-metal strip used in older relays and includes a cool down period. This can be seen on the T/C value.

(b) Low set overcurrent test with low set overcurrent enabled, set at 300% FLC and 10 seconds to trip:

- Inject at 300% of FLC, until relay trips, note exact rheostat position and switch off current.

- Set rheostat to zero and re-start quickly adjusting to 300% noting time taken, should be 10 seconds to trip. Current should be approx. 15A on secondary

(c) High set overcurrent is set at 1000%, this cannot be tested due to the high currents required for injection, e.g. 10 times secondary FLC. Therefore to test the O/C has to be reduced, proceed as follows

- High set overcurrent test with low set overcurrent enabled, set at 300% FLC and 10 seconds to trip:

- Disable Low Set Overcurrent and change HS Overcurrent value to 500%, leave at zero seconds trip time

- Quickly inject up to 25A secondary and note that the relay should instantly trip (if HS O/C is set to zero seconds)

- Repeat and record results for all 3 phases. Return to 1000% setting.

(d) Earth fault test with injection set adjusted to 0 – 10A output (to give easier control):

- Connect secondary injection leads, + to 28 and – to 27 on terminals at rear (or same terminals on front test block).

- Display scroll to "Io" screen to observe E/F current.

- Slowly adjust until E/F trips, note the secondary current value which should be approx. 20% of programmed FLC

- E/F set at 20% of FLC @ 850A = 170A, secondary value = 170A ÷(900/5) = 0.94A

- Slowly increase to 0.94A and observe trip. Leave rheostat in this position and re-inject noting the time taken should be within specified parameter setting.

Reset all parameters as per the provided parameter schedule.

6.5 3.3kV Circuit Primary Injection Testing

(a) Preface

A new pump/motor system was fitted that required less load for the same pump performance. The original had a running load of 100A, with 120:1 CT's fitted. With the new running load of 50A a 50:1 CT set was fitted.

The protection relay was a P&B Golds Mn4 1A rated. E/F set at 20%, therefore the secondary CT e/f value should trip at approximately 0.2A. A stabilising resistor also fitted at the original value of 179Ω but changed to 55Ω.

The general order and content to test following fitting of the CT's should be; Visual Inspection, CT polarity test, stability and sensitivity tests.

(b) Visual Inspection

After testing for dead the visual inspection can be carried out of the recent CT change. Visual and tactile inspections within the circuit busbar chamber as follows:

- Cleanliness of busbar chamber, no moisture, dust debris or tools left behind
- Correct CT's fitted and that they match
- Correct connection, tightness of secondary CT wiring and orientation of the CT positions

Figure 34 shows the actual CT's as fitted to the circuit busbars for the slurry system motor. The centre busbar has an extra CT; this is for the door mounted ammeter.

Figure 34: CT's fitted to Slurry System Circuit Busbars

(c) CT Polarity check

If new CT's are fitted it is necessary to check for correct polarity when installed, failure to do this will result in tripping.

CT's are identified as P1 = positive primary (e.g. the circuit conductor going through the CT), P2 = negative primary, S1 = positive secondary and S2 = secondary negative connection. Therefore the circuit conductor must flow from the circuit breaker outgoing through the CT (P1) and out the CT (P2) onto the motor. The picture below shows a CT polarity check on the bench as an example using a 6V battery and moving coil ammeter. See previous section on CT's for further information.

Figure 35: CT Polarity Bench Test

(d) Stability and sensitivity tests

The stability test is used to prove that the system remains stable during normal running and will not trip on earth fault i.e. there is no spill current. For the stability test a heavy short is placed across all three phases at the rear. In this case it was the point of primary portable earths as part of the HV isolation. The picture below shows the outgoing terminals with the cable box underneath. The portable earths connected on each phase and then joined together at a nearby earth point not shown on the picture.

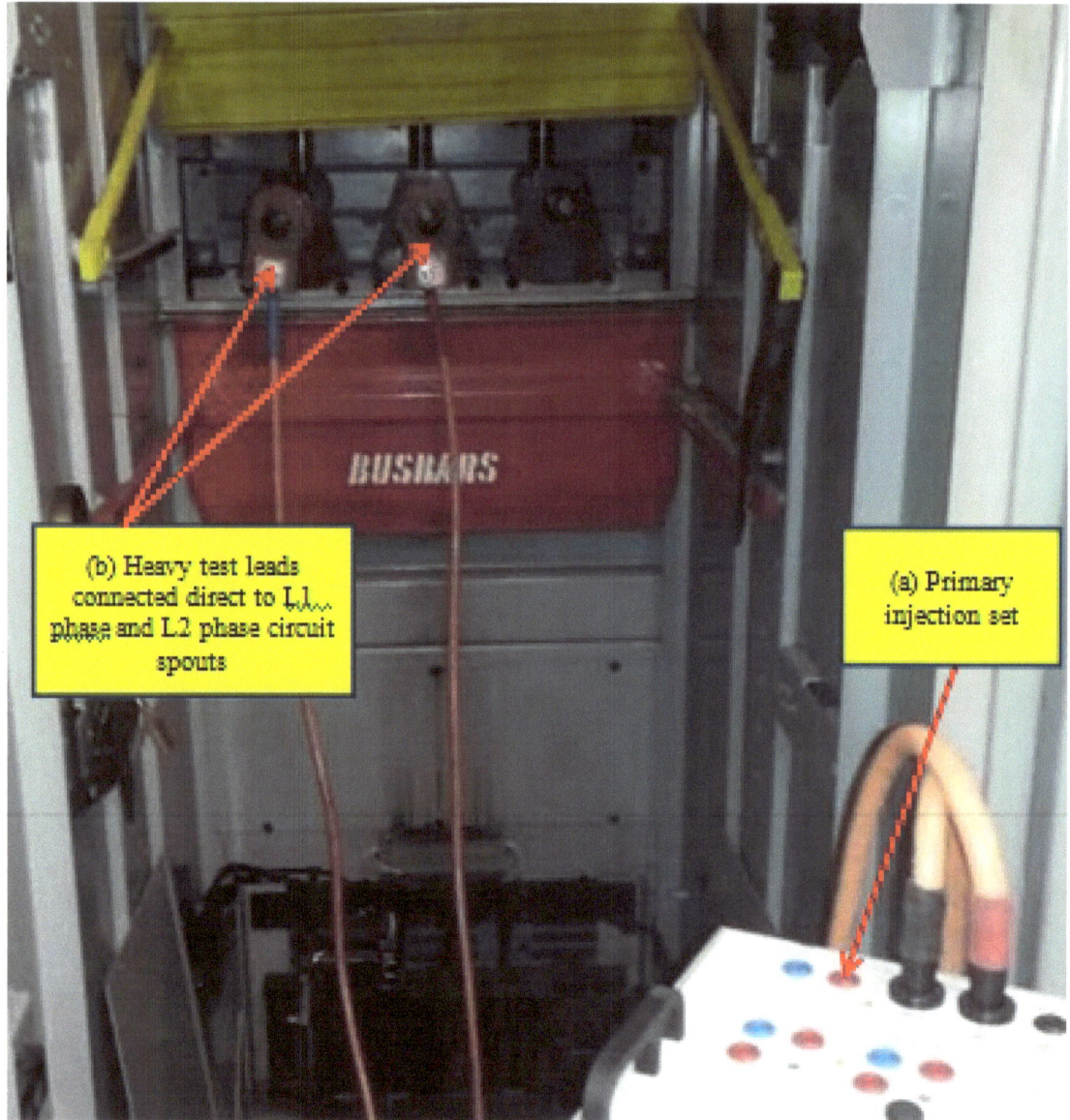

Figure 36: Stability check

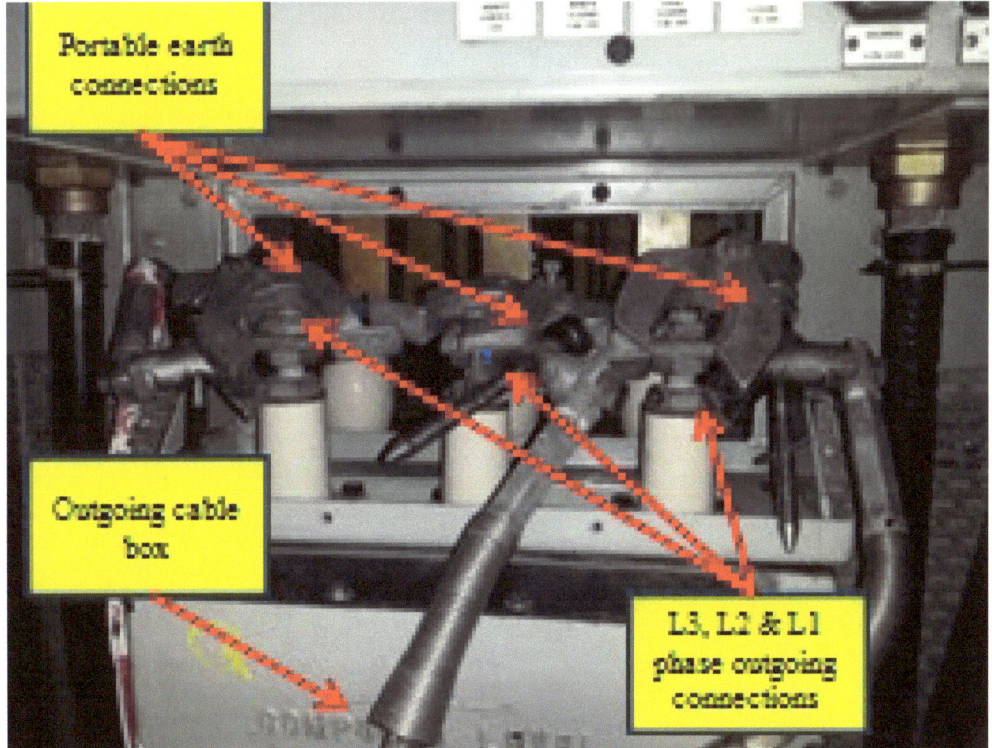

Portable earth
connections

Outgoing cable
box

L3, L2 & L1
phase outgoing
connections

Figure 37: Shorted outgoing terminals at rear of Switchgear

Stability test:

- Connect heavy current short across all 3 phases

- Connect primary injection set between L1& L2 at the circuit spouts

- Connect a calibrated clip on ammeter to the secondary wiring for L1 phase

Figure 38:Clip on ammeter position for checking secondary readings

- Inject 50A into the primary circuit (L1 to L2) and record the measured clip on ammeter value. Note it should be approximatley 1A due to 50:1 ratio. This test also proves correct CT ratio. Move clip on to L2, L3 and E/F secondary connections recording each value. At this stage there should be the same in L2, and neglible values in both L3 and E/F. The relay shouldn't trip.

Phase	Inj. Curr.	CT output at Relay (Amps)				Ammeter	Relay State
	Amps	L1	L2	L3	E/F	Amps	
L1-L2	50	1.010	1.060	0.002	1.85mA	50.000	STABLE
L1-L3	50	1.009	0.000	1.012	0.341mA	0.000	STABLE
L2-L3	50	1.8mA	1.006	1.008	2.04mA	50.000	STABLE
L1	15	0.268			0.196		TRIPPED
L2	15.4		0.262		0.190	15.000	TRIPPED
L3	15.3			0.269	0.197		TRIPPED

Table 6:Stability & sensitivity table

- The above test repeated for L1-L3 and L2-L3 with values recorded as per table 6

Sensitivity test:

- Although it is not needed the heavy short can be left on as per previous test
- Connect one heavy injection lead to L1 circuit spout and the other lead to a point on the L1 busbar just after the CT
- Inject 20% of FLC into L1 = 10A, the relay should trip but in this case it didn't due to the e/f stabilizing resistor still being set as per the previous installed motor ratings. The relay tripped at 22.3A injected with a stab value of 179Ω. The stab resistor for this relay was located inside the relay.
- With the stab resistor reduced to 55Ω a primary injected current of approx 15A tripped the earth fault relay at a secondary value of 0.196A.
- Although the primary current wasn't strictly 20% the secondary has to overcome the stab resitor and external resistance of wiring approximatley 8m length. The above test repeated for L2 & L3 wiith similar results.

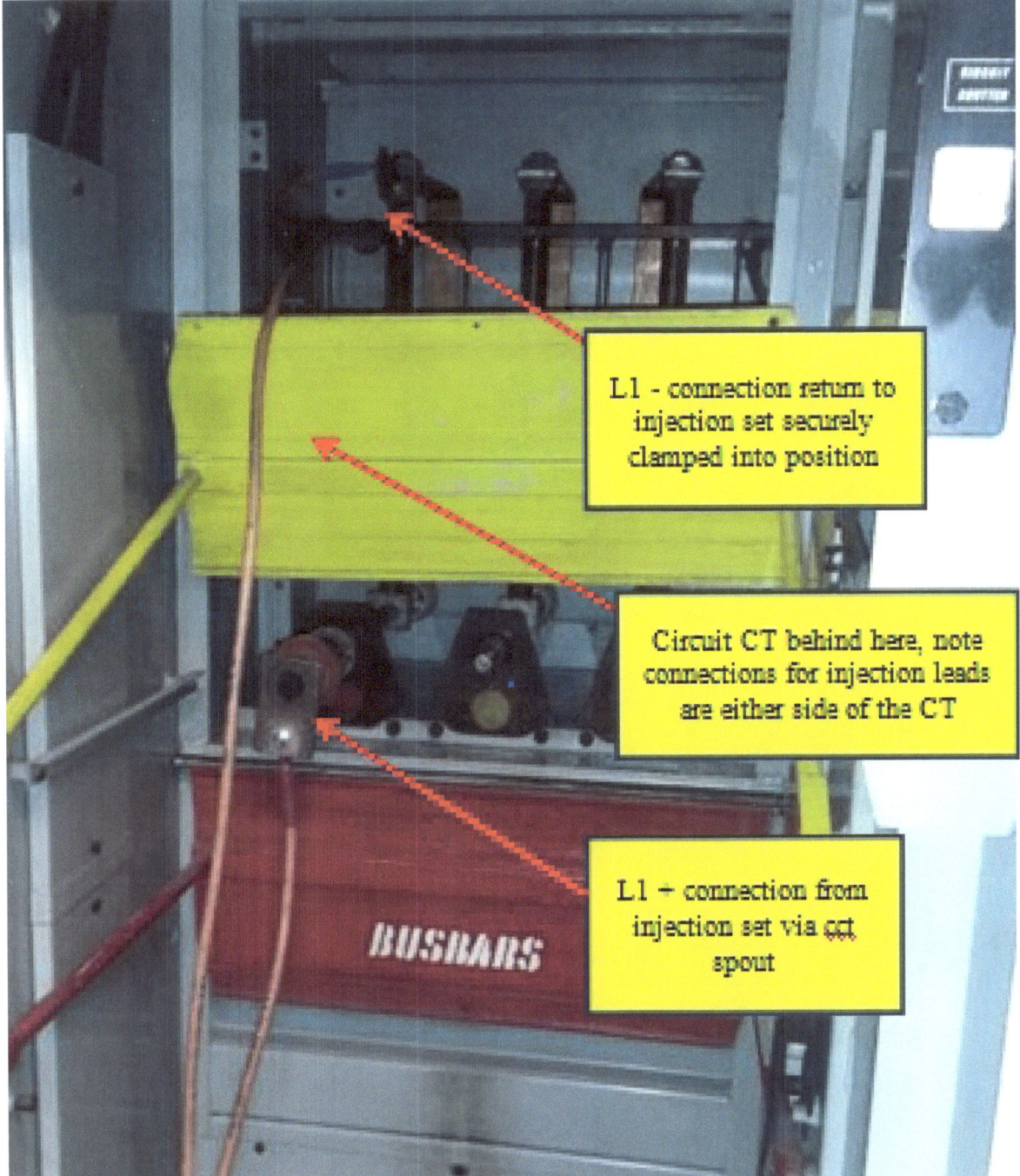

Figure 39: Sensitivity test at circuit spout

7.0 Further Reading

To enhance your understanding for this paper I've listed a few publications that have helped me during my career.

(a) IEEE Power Engineering Society, IEEE Recommended Practice For Testing Insulation Resistance of Rotating Machinery May 2008 Draft

(b) A Guide to Low Resistance Testing, Megger

(c) Practical Power System Protection, L.G. Hewitson et al

(d) Keeping Electrical Switchgear Safe (HSG230), HSE